# The Emerging Science of Water

Vladimir Voeikov and Konstantin Korotkov

Vladimir Voeikov and Konstantin Korotkov
The Emerging Science of Water

In this book, we would like to acquaint readers with the emerging new science of water. We were lucky enough to watch (and, as far as possible, to participate) in the development of this trend within the last 10 years. This book is intended to be user-friendly, reading like popular science. We mostly communicated using layman's language and avoided technical terms. We hope our readers will discover some ideas in this book that piques their interest.

Cover design by Oleg Bazhenov
Editor DARLENE

ISBN-13: 978-1973736820
ISBN-10: 1973736829

# Contents

# Introduction

*Our notions of physical reality can never be final.*
*We must always be ready to change these notions.*
Albert Einstein

Water is the basis of life; water is life itself.

One can live a long time without food, but it is impossible to live without water. Rivers, lakes, seas and oceans have fed half of humanity for centuries, and provided contact between peoples. Waterfalls and fountains attract people to the fascinating game of crystal jets. It seems, that after thousands of years of water use, we should now know all about this substance. Science has studied water for more than two hundred years. But, as it turns out, a lot of fundamental questions (that should arise if we look around with open eyes) have not even been considered in classical science.

Here are some of them:

Water evaporates. Sometimes we see evaporating water as a dense fog that gradually disperses and becomes invisible. This is no surprise. No questions arise. We take it for granted. But, why don't we ask why dissipating water vapors start to thicken to form clouds? Water molecules represent water vapors. A cloud is more than just liquid water. Is a cloud just a cloud? Reconsider how — a water vapor becoming thicker instead of dispersing—is counter intuitive. We don't give it a second thought, simply because we're accustomed to the phenomenon!

Some clouds turn into thunderclouds and become the source of a tremendous quantity of electricity manifested in the form of lightning. It is well known that water molecules may be ionized, but only if they are hit with UV-photons. It is obvious that density of UV radiation in the clouds is negligible. Why then, does so much water ionize in thunderclouds, allowing for the accumulation of the tremendous quantity of electrons needed for a lightning? Classical science cannot provide a definite answer to this question.

OK, however we understand the mechanism of electrical accumulation in thunderclouds and how lightning strikes the ground, we perceive the potential difference between the ground and the clouds. But then, why do discharges occur between the clouds? Where does the potential difference come from in these cases?

How does water rise up the trunk of a tree, sometimes at the height of multi-story building? Explanations using the capillary effect are inconsistent, which may be demonstrated by a simple calculation.

What about the healing power of "special" water treatments — with magnets, minerals including gemstones, through incantations and mantras? Are these superstitions and legends, or something real?

What is so special in the waters of springs and sacred sources?

How did the water appear on Earth?

Why is water detected on all planets (at least in the form of ice) and in space?

This list could go on, but classical science dismissively shrugs off such "trifles." But we know that often the new directions in science arise from consideration of just such little things. At the end of the XIXth century, the picture of the world was clear and understandable. Science was mainly formed and no one was able to imagine that the world was on the verge of quantum revolution.

The same situation is taking place now. At the beginning of the XXIst century, new previously unpredictable technologies are being developed, and a new scientific paradigm—a system of ideas about the world—matures gradually.

In biology and medicine, this marks the transition from the molecular to the quantum level; in physics, this marks the development of quantum non-locality and new ideas about the structure of the Universe; in the natural sciences, the inclusion of Consciousness in the world picture is marked. The new science of water is an essential part of all these areas.

### Let us introduce ourselves.

**Voeikov Vladimir**, Doctor of Biological Sciences, Professor. Professor of the Faculty of Biology, Lomonosov Moscow State University; author of over 400 scientific articles. Prigogine Gold medal 2013 winner. For many years, worked closely with Fritz Popp, then with Emilio Del Guidice.

**Konstantin Korotkov G.**, Doctor of Science professor. Professor of the Department of the St. Petersburg State University of Information Technologies, Mechanics and Optics; leading researcher of St. Petersburg Research Institute of Physical Culture and Sport; President of the International Union of Medical and Applied Bioelectrography; author of more than 400 scientific articles, 10 books, translated into many European languages.

In this book, we would like to acquaint readers with the emerging new science of water. We were lucky enough to watch (and, as far as possible, to participate) in the development of this trend within the last 10 years. Through long conversations with eminent scientists: Emilio Del Guidice, Gerald Pollack, Masarusaro Emoto, Jacques Benveniste, Fritz Popp, Rustum Roy, Harry Schwartz, and many others, this endeavor was facilitated by acquaintances. Thanks to them, we owe a debt of gratitude for a new vision of nature and life. Many of these scientists were in Russia at the annual international congress, "Science, Information and Spirit," held in St. Petersburg each year in early July (www.sis-congress.com).

We began writing this book a few years ago. But only after attending the Water Congress in Bulgaria in 2016 (www.waterconf.org), did all the ideas beautifully fall into place, like the pieces in a multicolored puzzle. For the next several months, we came together, polished these ideas, and are pleased to present the fruits of our labors to our dear readers.

This book is intended to be user-friendly, reading like popular science. We mostly communicated using layman's language and avoided technical terms. Sometimes, keeping it simple was not always possible, especially when discussing new scientific ideas. We did not include a detailed bibliography, but only the most essential—as data is easy to find on the Internet. We hope our readers will discover some ideas in this book that piques their interest. This is an excellent book to read in the evening, or while lying in bed, so as to draw one quickly into a dream state.

# Part I

# New Science of Water

# Special properties of water

*Water, thou hast no taste, no color, no odor; canst not be defined, art relished while ever mysterious. Not necessary to life, but rather life itself, thou fillest us with a gratification that exceeds the delight of the senses.*

Antoine de Saint-Exupery
*Wind, Sand, and Stars*, 1939

We can safely state, that biological life on Earth depends on the anomalous properties of water. Without it, life, as we know it, would have been impossible. Water possesses a whole lot of properties that set it apart from all other substances on Earth. The portal of the famous English physical chemist, Martin Chaplin, dedicated to water, contains the most complete information on water systems research (http://www1.lsbu.ac.uk/water/). He lists 72 "anomalous" properties of water, sharply distinguishing water from any other substances known to us. Every interested person is encouraged to study Chaplin's encyclopedia of water. But here, in this book, we review just a handful of properties on his list.

## The Anomaly of Density

The most well known anomalous property of water is related to its density being dependent upon temperature. The density of ice turns out to be less than the density of water, whereas, for all other substances, the density of crystals is greater than the density of liquid. This property of water makes it possible to preserve liquid water under a layer of ice.

If water behaved like other substances, ice would sink in liquid water and the water would freeze throughout the entire volume of the water body. In this case, gradually, the northern seas and oceans would turn into solid ice, which the summer sun would be unable to melt during the brief periods of seasonal warmth. If water behaved like other substances, the climate of the whole planet would become far more harsh and dry. By the way, we still do not know why the glacial periods happened or what part they played in the evolution of life on Earth.

Not only is ice less dense than liquid water, but also liquid water below 4ºC is less dense then at 4ºC, the temperature where water density is the highest. This is also an anomalous property of water.

The low density of ice, and the maximum density of water reached at 4ºC, results in the following:

(1) Prior to freezing, the whole volume of water—not only its surface—must be close to 0ºC,

(2) Freezing of rivers, lakes and oceans takes place from top to bottom, thus isolating the water from further freezing, reflecting the sunshine back into the space, which allows quick thawing,

(3) The density is regulated by thermal convection, which leads to seasonal changes in the temperature of deep waters.

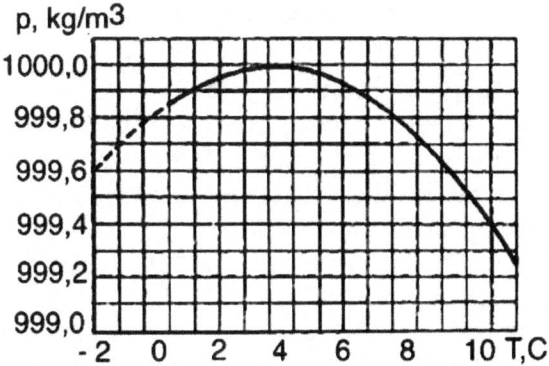
Temperature dependence of water density.

## The Anomaly of Surface Tension

Among the unusual properties of water, one of the most amazing is its extremely high surface tension of 0.073 N/m (20°C). Of all fluids, only mercury has a higher surface tension. Thanks to the high surface tension of water, some items being much more dense than water, may not sink in it immediately (e.g., a carefully laid steel needle). Many insects (water strider, spring-tails, and others) not only move on the surface of the water, but also take off from the surface into the air and "land" back on it as if it were firm ground. Moreover, living things have even adapted for use, the inner side of the water's surface. Mosquito larvae hang on it—with the help of non-wettable bristles—and small snails (mollusks and coils) crawl on its underwater surface in search of prey.

High surface tension allows water to take on a spherical shape in free fall or in a state of weightlessness: this geometric shape has minimal surface for a given volume.

The high surface tension of water is usually explained by different states of hydrogen bonds between adjacent water molecules present at the air-water interface and others being in bulk water and surrounded with water molecules from all the sides. However, it seems doubtful that hydrogen bonds between water molecules forming the surface monolayer, having the thickness of about 0,3 nm (0,00000003 mm), can increase so dramatically that this monolayer can hold a metal coin or a needle weighting fractions of a gram, or even more. Later, when we come to a new understanding of water properties and structure, we'll discover a much more realistic explanation relating to the surprisingly high surface tension of water.

## The Anomaly of Heat Capacity

The specific heat capacity of a substance is the amount of heat required to increase the temperature of one gram or one mole of a substance by 1 degree. There are several anomalies related to the heat capacity of water.

For the majority of substances, the heat capacity of a liquid increases only slightly (by no more than 10%) after the crystal melts. But water is entirely different in this respect. After the melting of ice, the heat capacity changes from 9 to 18 cal/ (mol·degree), i.e. with a twofold increase! No other substance manifests such a big change in heat capacity.

Water has the highest heat capacity among liquids, except for ammonia. A high value of heat capacity and high water content in organisms contribute to their thermal regulation, preventing local fluctuations of temperature. The high heat capacity of oceans and seas turns them into accumulators of heat. The temperature fluctuations of seas equal only one-third of the temperature fluctuations on dry land. This makes our climate more moderate and allows the ocean currents, such as the Gulf Stream, to transfer the heat of the tropics to the North-West of Europe.

Another striking anomaly related to the heat capacity of water, is that it has a minimal value between 36°C and 37°C. At temperatures below 36°C, the specific heat capacity of water decreases while temperature increases (unlike all other liquids) and above 37°C, as in all other liquids, it increases with heating.

Until now, there is no commonly accepted explanation for this phenomenon. What is striking is that this temperature range is characteristic of the normal body temperature of warm-blooded animals and human beings. Probably, this is not a coincidence.

## Other Special Properties of Water

Water is an excellent solvent—thanks to its polarity, high dielectric constant, and the small size of water molecules— especially for polar and ionic substances and salts. This is why it is so difficult to obtain genuinely pure water (e.g. with less than 5·10-9 of admixtures). Ice, on the other hand, is a very poor solvent, and why it is used in water purification (e.g. for de-gasifying) through using successive cycles of freezing and thawing.

Water has unique hydration properties with respect to biological macromolecules (especially proteins and nucleic acids), thereby defining their three-dimensional structures and, consequently, their functions in a solution. Water can be ionized and allows an easy exchange of protons between the molecules, thus contributing to the ion interactions in biology.

The opposite properties of cold and hot temperatures upon water are truly fascinating. When heated, cold water is compressed and becomes less compressible. As its refractive index increases, the speed of sound in water increases, gases become less soluble and water becomes more easily heated while conducting heat better.

Conversely, when hot water is heated, it expands and becomes more compressible. As its refractive index decreases, the speed of sound in water also drops, gases become more soluble. Also, water becomes increasingly hard to heat and its conductivity keeps falling. Under increased pressure, the molecules of cold water move faster, while molecules of hot water slow down.

Hot water freezes faster than cold water. Ice melts upon compression—with the exception of high-pressure areas, where water freezes upon compression. No other substance can be so readily seen in all three forms at the same time: solid, liquid and gaseous. Scientists identify 14 different forms of ice, depending on its pressure and temperature.

For hundreds of years, water's properties seemed mysterious to mankind. In the last century, many fantastic hypotheses were formulated regarding the properties of water. Only recently, have we been able to even attempt explaining those anomalies. This is thanks to the appearance of a new class of experimental data, along with the evolution of advanced ideas about the structure of water.

All these issues pertain to state-of-the-art physics. They are still very controversial and lie far beyond the matters discussed in this book. However, in many situations, even without a profound understanding of the details, it is enough to know that more knowledge regarding water's anomalies exists and is being developed.

For instance, we all know that sunspots appear on the Sun, but few people are familiar with corresponding modern theories.

## Magnetic and Electrical Properties of Water

The technical effects of magnetized water are well known, documented, and put into practice. Magnetized water reduces the buildup of deposits of hard water in pipes and technical installations—including the lime scale formed during boiling—and loosens calcium sediments that are subsequently washed out. Curiously, these effects remain in force for a long time after the magnetic field disappears. Also, a lot of empirical data has been published describing the effects of magnetized water on plant growth, such as the increased germinating capacity of treated seeds, as well as changes in physical properties of "treated" water.

Although the effects of water being treated with magnetic fields are becoming a widely used industry norm, until recently, academic scientists rejected the possibility that weak and moderate magnetic fields can, more or less, change water properties in a stable way. Such a claim is based on the conventional model of water—the so-called flickering cluster model—in which the rotational relaxation of water molecules in water is known to be about 10-11 s.

On the other hand, the time needed for the preservation of the "memory" of the magnetic treatment of water after its passage through a magnetic field, should reach a minimum of several minutes. But, according to the data of many authors, water treated with magnetic (and electromagnetic) fields may actually reach an exposure of many hours and even days. Most authors limited their explanation of these phenomena. Rather than focus on the water alone, the results, they claimed, were just "the effects of admixtures present in the water."

Fortunately, in the last few years, several serious studies demonstrate the influence of weak and super-weak constant and variable magnetic fields have upon aqueous systems (that do not contain admixtures sensitive to the action of magnetic fields). A large series of studies on the influence of weak and super-weak constant and variable magnetic fields was conducted at the Institute of Cell Biophysics in Puschino, near Moscow. These studies were convincingly demonstrated; the results have been published; and the number of similar studies is constantly on the rise. The studies provide definite results regarding the effects of magnetic fields on biochemical processes—both through direct influence of the magnetic fields, and through their influence upon water solutions in which biological organisms were developed. For example, a combination of a weak static magnetic field (42 $\mu$T) and a low-frequency variable magnetic field (40 nT, 3-5 Hz) can change the intensity of the fluorescence of certain proteins and their functional activity. Exposure of *Dugesia tigrina planarians* to a combined magnetic field increased the intensity of their motor activity. Even more surprising, is the fact that water treated with a magnetic field transferred this effect to untreated planarians[1].

The use of magnetic water for three months resulted in a statistically significant reduction in dental plaque by 44% in a group of volunteers, in comparison with the control group[2]. Significant effects for growing chickens were detected using magnetic water[3]. The permanent magnet, created by a 600 Gauss strength magnetic field, was installed on the pipe through which flowed the water for the chicken feeder. One month later, 50 chicks in the experimental group showed an increase in weight by 200 g compared with 50 chicks in the control group[4].

---

[1] Novikov V. Biological effects of weak magnetic fields. PhD Thesis. Pushino 2005 (in Russian)

[2] Johnson, KE; Sanders, JJ; Gellin, RG; Palesch, YY (1998). The effectiveness of a magnetized water oral irrigator on plaque, calculus and gingival health. Journal of Clinical Periodontology 25 (4): 316–21.

[3] M. Gholizadeh, H. Arabshahi, M.R. Saeidi and B. Mahdavi. The Effect of Magnetic Water on Growth and Quality Improvement of Poultry. Middle-East Journal of Scientific Research 3 (3): 140–144, 2008.

[4] www.life-sources.com/pages/The-Health-Benefits-of-Magnetic-Water.html

Chickens that drank magnetized water were healthier, had fewer inflammations and had the best ratio of meat to fat. The authors also note that in the pipe with the installed magnet, deposits did not form, as opposed to conventional tubes. In Israel, the use of magnetic water for 85 cows increased their milk yield by a liter of milk a day. The cows also appeared healthier and their calves gained more weight than in the control group. Magnetized water has a significant effect upon plant growth. A statistically significant increase in seed germination of rice, beans and tomato has been demonstrated[5]. These seeds also germinated 2-3 days before the seeds within the control group. Carefully controlled experiments conducted in Australian greenhouses showed that magnetic water has a stimulating effect upon some plants (yield increases by 23% for celery and by 6-8% for beans), but had no noticeable effect on others. The effect is strongly dependent on the type of water used[6]. The number of similar examples can be substantially increased as a lot more information can be found on the Internet[7].

**If electromagnetic effects can really influence the degree of water structuring, they obviously can influence our health, because we consist of over 70% water. Biological effects of electromagnetic waves of Super High Frequency and Extremely High Frequency bands, which are very popular in Russian medicine, and the effects of successful healers prove the effectiveness of the influence of generated fields upon water systems.**

Finally, by the end of the last century, several groups of scientists managed to discover conditions under which the influence of magnetic fields upon water was stable and reproducible. It turns out that the influence must always be dynamic. This means the water must flow through the field area at a sufficiently high rate of speed, and the field must be orthogonal.

---

[5] Carbonell, M.V., Martinez E., Amaya J.M.. Stimulation of germination in rice by a static magnetic field. Electro-Magnetobiol, 2000. 19(1): 121–128

[6] Maheshwari B.L., Greval H.S. Magnetic treatment of irrigation water: Its effects on vegetable crop yield and water productivity. Agricultural Water Management 96, 1229–1236, 2009.

[7] www.fractalwater.com/research/magnetic-water-technology-research/

The best results were achieved with several constant magnets positioned some distance from each other on a pipe with water. There seems to be some kind of resonance created during the movement of a water stream along a path of individual magnets[8].

Therefore, a large amount of experimental data has been collected, proving that weak and super-weak magnetic fields have reproducible effects upon water solutions, and that these effects are most prominent when static and variable magnetic fields are used simultaneously. As was noted above, the greatest effects of water magnetization are manifested for water with high salt content. The dissolved salts induce water clusters that are easily polarized in the presence of electric and magnetic fields. Polarization facilitates the binding of additional molecules of the dissolved substance, and water molecules take all free valence electrons. Thus, after efficient treatment of the solution, free molecules of the dissolved substance become virtually absent in the solution, and they do not precipitate out upon the walls of pipes and vessels. Moreover, the polarized clusters capture the molecules that have already precipitated on the walls, thus constituting an efficient cleaning process. The process also depends on the speed of movement of the liquid. In accordance with the Lorentz Force Law, an electromotive force (EMF) is induced in a conductor passing through a magnetic field, said force facilitating the polarization of water clusters. The optimum condition is determined in an empirical manner depending on the electrical conduction and hardness of water. On the other hand, reliable data has recently appeared that suggests magnetic and electromagnetic fields can significantly influence properties of highly diluted aqueous solutions and include aqueous systems. This defines homeopathic solutions. Their effects cannot be explained in the framework of conventional water models. However, new explanations based on new insights about aqueous systems will be addressed next.

---

[8] Busch K.W., Busch M.A. Laboratory studies on magnetic water treatment and their relationship to a possible mechanism for scale reduction. Desalination 109, 131-148, 1997.
Kronberg K.J. Magnetic water treatment de-mistified. July 1999

# Water in Space

*The surface of the Earth is the shore of the cosmic ocean.*
*On this shore, we've learned cosmic ocean.*
*On this shore, we've learned most of what we know.*
Carl Sagan (1934-1996)

Recent data has shown that water is the third most abundant substance in the Universe, after hydrogen and helium. Basically, it exists in molecular form in gas clouds, but also in the form of ice in asteroids and planets. Enormous accumulations of water in deep space have been discovered in 2011. Water, equivalent to 140 trillion times all the water in the world's ocean, surrounds a huge, feeding black hole— called a quasar—more than 12 billion light-years away. "The environment around this quasar is very unique in that it's producing this huge mass of water," said Matt Bradford, a scientist at NASA's Jet Propulsion Laboratory in Pasadena, CA., "It's another demonstration that water is pervasive throughout the universe, even at the very earliest times." Bradford leads one of the teams that made the discovery. His team's research is partially funded by NASA and appears in the Astrophysical Journal, *Letters*.

The volume of water in such clusters is trillions of times greater than the volume of all the Earth's water resources. The assumption by astrophysicists is that a huge amount of ice lies outside the boundaries of the solar system in the "Oort cloud." This cloud represents the remnants of the process of forming our solar system. It consists of comets with orbits that suggest they belong to our planetary system. The nuclei of comets are composed of huge blocks of ice and snow, interspersed with comic dust particles. The number of these comets may reach several trillion, and the total weight is tens or hundreds of times the mass of Earth. In our Solar system, water may be the first compound in abundance. According to contemporary estimates, the quantity of water and—though it may seem astonishing—liquid water, outside the Earth may constitute 25-50 times more than is present in the oceans on Earth. The first case of detection of liquid water in space was on Enceladus, a satellite of Saturn.

In 2004, the cosmic station "Cassini" reached the Saturn system and had found signs of water on the surface of Enceladus. Since the surface of Enceladus is white, it is believed that the planet is covered with a layer of water ice. Under the ice shell may be pools of water in a liquid state, which jets to the surface as geyser flows and freezes. "Cassini" calculated the fountains of water to be many hundreds of kilometers in height, beating four cracks near the south pole of the planet. The temperature of this water may be about 0°C. The ambient temperature of the surface of Enceladus is minus 180 - 190°C, but inside this small planet, water should be warm. The question, "What force can heat water under the icy crust," is still unknown to scientists. But it is certainly not the Sun. "We were able to find an environment in which life could have arisen in a rather unexpected place of the solar system. But of this, we cannot say with absolute certainty, at least as long as we do not get the opportunity to go there," said Carolyn Porco, the representative of the Institute for the Study of Space in Boulder, Co.

The "Cassini" station had also explored the largest satellites of Saturn—Titan and Rhea. More than half the surface of Titan consists of frozen water. NASA experts say that if we can find more of such planets with the presence of water, it is quite possible that on the icy satellites of other planets and stars there will be suitable conditions for life to emerge. Recently, there have been indications that the water content in the earth's crust and deeper, may be much greater than was previously thought and can possess many times the water content on the earth's surface. In the light of recent findings, the estimate of water content needs to be updated—here and on other planets. Huge masses of ice, thousands of miles thick, are concentrated in the depths within the giant planets like Saturn and Jupiter. Europa, Jupiter's moon, is covered with ice, under which–according to scientists—this water is in its liquid state. A large number of ice chunks, likewise, can be found on the surfaces of planets, like Neptune and Uranus. Triton, the largest moon of Neptune, is also mainly composed of water ice.

Water, has also been discovered on our Moon. According to the data transmitted by the radar Mini-SAR, (established on the Indian lunar apparatus Chandrayaan-1), just on the Moon's north pole region, it discovered at least 600 million tons of water—much of which is in the form of blocks of ice—resting on the bottom of lunar craters. Water has been found in more than 40 craters with diameters ranging from 2 to 15 km. Now scientists have no doubt that this ice—is water ice.

# The New Science of Water

*Thousands have lived without love,
not one without water.*
W. H. Auden, English poet
(1907-1973)

We began this book by mentioning that water has properties vastly different, physically and chemically, from almost all other liquids. From the standpoint of classical physics and chemistry, water would be considered as an "abnormal" substance. In this case, the term "abnormality" implies that other fluids (or at least the vast majority of them) are "normal," at least they share similar properties with each other, which also differ from the properties of water.

"Anomalous" water—a very inconvenient coin for the "normal" academic scientist to use because from the standard models of classical physics (thermodynamics, statistical physics, electrostatics and electromagnetism, which form the basis of physical chemistry and biology), these "anomalies" are not followed. But the main advantage of what we call "science," is the possibility of a confident prediction of properties of substances or processes based on "accurate scientific knowledge."

Attempts to explain the "anomalous" properties of water in the framework of classical concepts of physics and chemistry were persistently undertaken during the last century. Particular attention was paid to the fairly unique feature of water molecules: the unequal distribution of the electron density in this small molecule, a lack of negative charge on the hydrogen atoms and the excess one on the oxygen atom.

This allows water molecules their easy electrical polarizing-ability and ensures their ability to form the so-called hydrogen bonds—directional interaction of electrostatic nature, but weaker than electrostatic interactions between the negative and positive charges. Each water molecule can theoretically form up to 4 hydrogen bonds with neighboring molecules, but in reality, the bonds are fewer. Bonds are directed, but they can be bent, and the group of water molecules may, at any time, form a unique pattern. Many theorists try to explain this behavior as the special properties of water "anomalies." A detailed catalog and analysis of various theoretical models can be found on the Martin Chaplin portal (http://www1.lsbu.ac.uk/water/).

What are some model assumptions about the structure of water, based on the concepts of classical physics and chemistry? These concepts cannot, in any way, explain a number of phenomena connected with the properties of water, some of which are regarded as mystical. Such properties include, for example, the retention of biological activity in the water samples obtained by multiple dilutions of an aqueous solution of biologically active substances to such an extent that within the solution there cannot theoretically be any active molecule. This happens to be a principle of homeopathy, founded more than two hundred years ago, but still rejected by official science. However, the practice of homeopathy is not only backed by numerous experimental facts and observations made by numerous doctors, but it is also widely used in practice.

In this section, we began with the discussion of the most striking (but not the only) studies relating to these absolutely "anomalous" (and from the point of view of scientific fundamentalists— "paranormal") properties of water. Now, we'll show that modern physics, in contrast with classic physics, not only "allows" water to behave in a "strange" way, but also requires it to display such a behavior.

# The Drama of Jacques Benveniste

*I can calculate the motion of heavenly bodies,*
*but not the madness of people.*
Isaak Newton (1643-1727)

In 1983, the homeopath Bernard Poitevin asked the famous French immunologist, Jacques Benveniste, to investigate the biological effects of some homeopathic solutions he developed at his laboratory. Homeopathic medicines are solutions of the active substance diluted to such an extent that they cannot contain molecules of the starting substance. According to the laws of chemistry, reaction of the biological system to the preparation must decrease and eventually disappear with decreasing concentrations of the active molecules. Benveniste, a brilliant scholar of classic science, was convinced that homeopathic remedies could not have an effect on a biological object.

Nevertheless, he agreed to create such experiments, perhaps to prove, yet again, the failure of homeopathy. But to his great surprise, the results he obtained were in complete contradiction with the generally accepted views.

As a test system, Benveniste and his colleagues used a sensitive biological model in their laboratory, which is widely used in other immunological laboratories. It is known that basophils (one type of immune human cells) rapidly react to various biologically active substances; in particular, they have specific interactions with antibodies. The reaction of basophils is called a de-granulation reaction, and it's reliably detected by the change of cells colored by specific dyes.

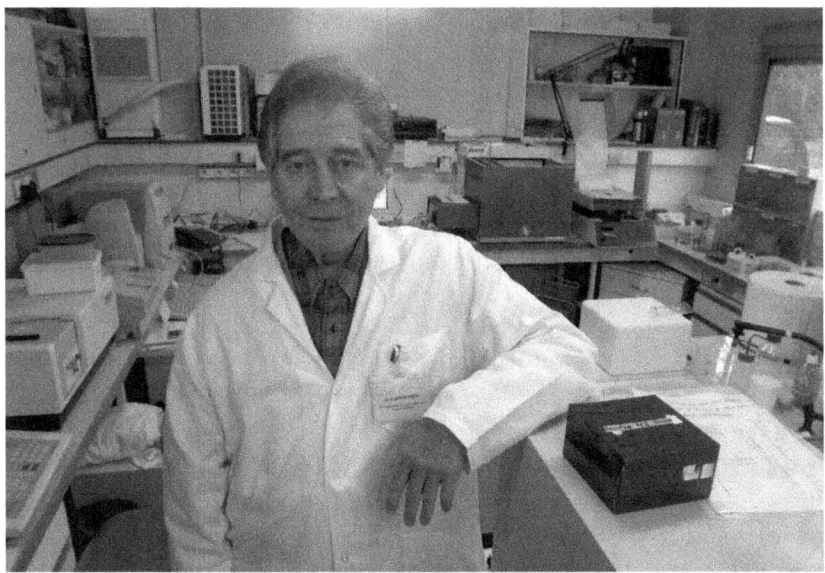

Jacques Benveniste

Researchers have taken specific antibodies against proteins basophils and the reaction was tested on series of dilutions in which the concentration of antibodies was successively reduced tenfold each time. Calculations show, that after the 20th such dilution (and higher), the probability of detection of at least one antibody molecule in the sample was negligible. Experiments have shown that in the beginning with a reduced concentration of active substances, the effect was, as expected, declining. However with further dilution it rose again, and at subsequent dilutions varied in wave pattern, which was quite unusual (see Fig. 1).

Such a regular change in the biological activity of "solutions" of antibodies was observed up to 10-120 dilutions. Essential to the success of the experiment was, that after each "breeding" session, the new "solution" was repeatedly shaken. This method repeats the breeding technique of preparation of homeopathic remedies developed by Hahnemann. Benveniste suggested that the transfer of biological information was due to the fact that it was "imprinted" in the structure of water. Essentially, he confirmed that there was a "memory of water."

Researchers submitted articles describing their studies, aiming to be published in the most prestigious scientific community journal, *Nature*. The editorial staff expressed concern that the publication of this material will allow homeopathy practitioners to claim that homeopathy is scientifically proven, even if the Benveniste statements will be subsequently denied. The editor of *Nature*, John Maddox said: "Our mind is not as so much closed as not ready to change the idea of how modern science is constructed." After a lengthy review process, in which the referees insisted on seeing evidence that the effect could be duplicated in three other independent laboratories, *Nature* finally published the paper[9].

However, what was quite unusual for the editorial policy, John Maddox, prefaced it with an editorial comment entitled *'When to believe the unbelievable'*, which admitted, "There is no objective explanation of these observations."[10]

Maddox claimed that since the published results contradict the laws of physics, the editors would check how expertly the experiments were conducted in the Benveniste laboratory. Indeed, soon to the French lab came a commission composed of the Maddox—a physicist by training, James Randi—a magician who, by his own statement, specializes in exposing charlatans, and Walter Stewart, a physicist by training. Stewart was not the experimenter, but a freelance inspector of the National Institutes of Health.

The group came to the Benveniste laboratory and demanded that the experiment be repeated in their presence. The first series of experiments were carried out in accordance with the protocol described in the Benveniste article. These data coincide with the published data. However, Maddox and his colleagues demanded a repetition of the experiments with their own protocol (let us recall that none of them were biologist experimenters). Also, let's note, that in their experiment, basophils were used, which are not sensitive even to high doses of the antibody. Naturally, the result was negative. Despite the fact that the Benveniste article in *Nature* was based on the results of more than a hundred experiments with significantly positive results, one failure was enough to be

---

[9] E. Dayenas, F. Beauvais, J. Amara, M. Oberbaum, B. Robinzon, A. Miadonna, A. Tedeschi, B. Po- meranz, P. Fortner, P. Belon, J. Sainte-Laudy, B. Poitevin J. Benveniste. Human basophil degranula- tion triggered by very dilute antiserum against IgE. NATURE VOL. 333, 30 1988, p 816-818

[10] Maddox J, Randi J, Stewart WW. 'High-dilution' experiments a delusion. Nature 1988;334:287–90

published in a subsequent issue of *Nature,* reporting Benveniste's lab as "fraudbuster." The final part of the report read: "There is no reason to suggest that the antibody in high dilutions retain their biological activity. The hypothesis, that water has a memory of past solutions, is as useless as far-fetched."

Reaction to the Benveniste article in the academic world was not merely negative, it was literally life destroying. Benveniste was almost charged with fraud. Before the scandal that erupted after the publication of an article in *Nature,* it should be noted that, Benveniste enjoyed an impeccable reputation and was recognized as one of the world leaders in the field of allergy and immunology. He has discovered an important factor in inflammatory processes—the platelet-activating factor (PAF). Incidentally, this factor acts in vivo in such low concentrations, that it casts doubts on the accepted mechanisms of chemical reactions. This fact somehow contributed to the fact that Benveniste began working with ultra-low doses.

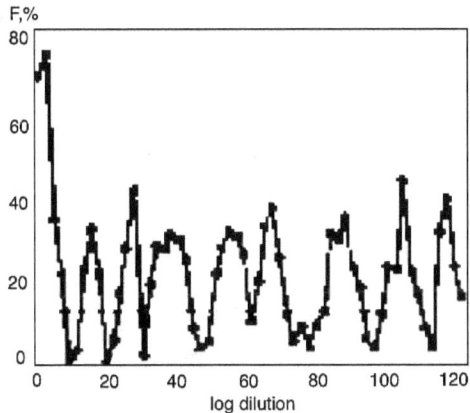

*Fig. 1. The dependence of the degree of basophil degranulation (%, y-axis) on the degree of dilution of antibodies (x-axis, the logarithm of dilution).*

Unfortunately, his former merits failed to protect the "heretic" from the wrath of science's fundamentalists. He lost his funding and then his position at the Pasteur Institute, where for many years he headed the laboratory. Such an attitude toward Benveniste and his research has been maintained in academic circles ever since. In numerous Internet publications, science fundamentalists continue to nay say the phenomena reported by Benveniste. Contrary to the "well-established laws of science" they claim his results reflect his mistakes and incompetence, and his results should never have been confirmed and were experimentally refuted. To see the falsity of such allegations is sufficient to search databases of scientific articles cited in the original Benveniste article in *Nature*. In other major scientific journals, articles were published in which Benveniste's results were entirely verified[11].

Despite the obstruction and a boycott by members of academia, not only did Benveniste not stop studying the biological effect of ultra-low doses (so now in science everything associated with homeopathic dilutions is called) of biologically active substances (BAS), he engaged more daring research. He became the progenitor of a new scientific direction—the transport of BAS activity via electronic networks to clean water, which as a result, acquires properties of the initial substance.

Already in the *Nature* article, he suggested that the high biological activity of the dilutions in the absence of antibody molecules might be caused by water. Benveniste suggested that water in the process of dilution, accompanied with potentiation (shaking), acquires the properties of electromagnetic "matrix" for bio-molecules, which has unique electromagnetic properties (frequency). Even though, at the end of his article, Benveniste recognized that this hypothesis was not yet substantiated, he decided to put the "insane" experiment into action. Using an electronic amplifier, Beneveniste transferred the electromagnetic information of a given substance, to clean water via an electrical circuit. Even Benveniste's many supporters initially reacted to this challenge with skepticism. This reminds us of the skeptical reactions ordinary people display when first exposed to new ideas, like the transmission of the human voice over electrical wiring or flying with machines heavier than air.

The experiments began in 1992. A sealed tube containing a

[11] Belon, J. Cumps and M. Ennis *et al.*, Inhibition of human basophil degranulation by successive histamine dilutions: results of a European multi-centre trial, *Inflamm Res (Suppl 1)* **48** (1999), pp. S17–S18.

Brown V. Ennis M. Flow-cytometric analysis of basophil activation: inhibition by histamine at conventional and homeopathic concentrations. Inflamm Res 2001; 50(Suppl 2):S47–S48 solution of biologically active substances (eg, stimulant of the respiratory burst of neutrophils) was placed in a conductive coil. The coil was connected via an electronic amplifier with another coil in which a biological detector was placed, such as a test tube with a suspension of neutrophils. The current was turned on and in a few minutes, there was a biological response—the intensity of the respiratory burst was registered. The reaction efficiency was *evaluated* by increasing the respiration rate of the cell suspension as compared to what it was before. If a sealed vial with pure water was placed in the receiving coil and treated for 15 minutes with a signal from a BAS solution, water acquired the activity of the sending BAS[12].

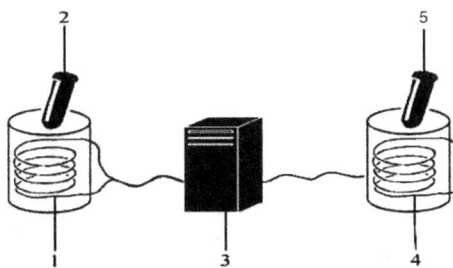

Fig. 2. *Device for carrying biological signal by the electronic circuit. 1 - the coil for receiving a signal; 2 - test tube with a solution of biologically active substances or control water; 3 - amplifier (PC); 4 - coil for transmitting a signal; 5 - tube with suspension cells or clean water.*

Since the electronic amplifier placed between the coils was a conventional audio amplifier, Benveniste was able to record signals from the amplifier to the hard disk of a computer. The information was recorded in *.wav format on electronic media (eg, CD-ROM) and could be passed to any distance[13]. For example, pathogenic

---

[12] Y. Thomas, M. Schiff, M.H. Litime, L. Belkadi, J. Benveniste. Direct transmission to cells of a molecular signal (phorbol myristate acetate, PMA) via an electronic device. FASEB Journal (9: A227 (abs) 1995

[13] Y. Thomas, L. Kahhak and J. Aissa, The physical nature of the biological signal, a puzzling phenomenon: the critical role of Jacques Benveniste. In: G.H. Pollack, I.L. Cameron and D.N. Wheatley, Editors, Water and the Cell, Springer, Dordrecht (2006), 325–340

strain of E. coli bacteria (causing aggregation of latex particles to which the antibody was bound in the cells) was taken. Recorded from the bacterial suspension, this information was "played" to the tube with water, resulting in the latter acquiring the antigenic properties of the original cell suspension. Effects were always reliable, even though the magnitude of the effect varied on different days. This technology allows the detection of an immunogenic substance at any distance and opens up many important opportunities for tele-diagnosis, such as used with infectious diseases.

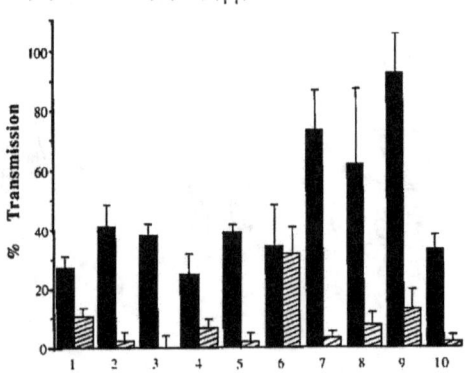

Fig 3. *The results of 10 independent experiments on the transfer of signal from the active stimulation of neutrophils solution (black bars) and the solvent (striped bars). Significant difference between the experiment and control were found in 9 out of 10 in this series of experiments.*

Benveniste experimented publicly. He invited independent researchers to reproduce his results. During his lifetime, Benveniste did not receive scientific recognition for his groundbreaking work. But at the same time, for doctors involved in alternative medicine—homeopathy, color therapy, electromagnetic therapy and related areas—his name was widely known and respected. He was regularly invited to speak at international congresses. Their acceptance did not equal the "caliber" of science, something Benveniste was used to during the early years of his academic career. Jacques Benveniste died in 2004. Sadly, he did not last long enough to witness the rise of a new scientific field related to the structural properties of water.

Benveniste was not the first and the only researcher who has demonstrated the possibility of transmitting biological information at the distance. Such experiments have been reported before and a considerable portion of them was carried out in the USSR and Russia. The difference is that Benveniste's work most strictly met the rigors of scientific criteria for biological experiments.

The reasons behind science's rejection of Benveniste's work and the subsequent personal attacks against him are discussed in detail in the Michel Schiff's book *"Memory of Water."* This book is the result of an in-depth study of scientific and human drama, which had become the storyline in the life of Jacques Benveniste. The author and many other observers rightly compare Benveniste drama with Galileo drama. In Galileo's case, not only was the Catholic Church responsible, so too, were his fellow scientists. Among many subjective factors related to human nature, there is one that may be considered objective.

The results obtained by Benveniste contradicted long-held beliefs of physicians, biologists and chemists, which for more than a century had served as the basis of biology and chemistry. By the end of the twentieth century, these beliefs allowed impressive achievements in the understanding the phenomena of life and progress in the practical use of this knowledge. In turn, these ideas were based on the seemingly immutable, firmly established laws of physics—or more precisely, on the laws of classical mechanics and electrodynamics. Maddox, in fact, insisted that the results reported by Benveniste, have no right to exist, because they contradict the truth as established by science. And most scientists agreed then and still are in agreement with Maddock's position.

But the question arises—whether theoretical concepts (which we often call the immutable laws of nature) developed in classical physics of the XIXth century are enough to explain all the new phenomena discovered in biology over the past half century, given the fact that the fundamental physical principles have been dramatically transformed in the last century.

Think of quantum theory, the thermodynamics of open systems, the physics of non-equilibrium and non-linear processes. At the same time, Benveniste, having understood all that, knew his discovered phenomenon was real. He only lacked a rational scientific explanation, and this explanation can be sought in the field of quantum physics. In the early 1990s, he turned to the Italian theoretical physicists, Giuliano Preparata and Emilio Del Giuidice, for help. Based on the firmly established laws of quantum field theory and quantum electrodynamics, these two scientists justified this fundamentally new model of water. Not only does the model follow the nature of water as discovered by the Benveniste phenomena, but also other water-related phenomena that before seemed impossible within the framework of classical ideas. We will discuss their ideas in detail, as these represent the most fundamental approach to a modern understanding the nature of water.

Thus, Benveniste became the progenitor of a new scientific direction—transfer of Biologically Active Substances (BAS) properties by electronic networks to clean water, which, as a result, acquired properties of the starting substance. Benveniste experimented publicly, often involving independent researchers, who successfully reproduced his results.

# Neglected Facts about High Dilutions

Benveniste was not the first and not the only researcher who has shown the ability to transmit biological information without direct interaction of biologically active molecules with a particular cellular receptor. Reports of similar experiments were published before, and a considerable part of them was carried out in the USSR and Russia.

We do not aim to provide an overview of the thousands of papers published in this area—it is easy enough to find them on the Internet[14]. From the vast array of studies that exist, we choose to mainly focus on the most fundamental as carried out by Russian scientists because much of their research has undeniable scientific value, they are not known to the scientific community plus a broader familiarity with them will be an further impetus to develop new scientific concepts.

One of the first fundamental studies regarding the effect of extremely highly diluted solutions of BAS on living systems was conducted back in the last century in the twenties. Nikolai P. Kravkov (1865-1924) was a renowned scholar who has made a recognized contribution to pharmacology, physiology and clinical medicine. He was an academician of the Imperial Military Medical Academy (1914), corresponding member of the Russian Academy of Sciences (1920). Official Soviet medicine recognized him as the founder of domestic pharmacology. His textbook *"Fundamentals of Pharmacology"* withstood 14 editions and became the base for several generations of physicians.

By the end of its productive scientific life, Kravkov undertook a large study on the possibility of reducing the dose of biologically active substances—like adrenaline, histamine, nicotine, strychnine, quinine, and ether—which in high concentrations act as poisons. Would it still be possible to register response to them by living systems when the dose is reduced?

Kravkov investigated the effects of these substances. The method he used, now incorporated into the history of medicine as "Kravkov – Pisemsky" Method, measures the blood flow velocity in

---

[14] http://hydrometeorology.ws/str42.html.International Journal of High Dilution Research

an isolated rabbit's ear. Administered through the vascular system of the rabbit's ear was saline solution to which was added a given test drug. The effect of the drug was measured by the liquid flow rate through the vessels.

*Doctor of medicine N.P. Kravkov*

Kravkov investigated the effects of these substances. The method he used, now incorporated into the history of medicine as "Kravkov – Pisemsky" Method, measures the blood flow velocity in an isolated rabbit's ear. Administered through the vascular system of the rabbit's ear was saline solution to which was added a given test drug. The effect of the drug was measured by the liquid flow rate through the vessels. Substances, which normally have vasoconstrictor action concentrations (adrenaline, histamine, etc.), at high dilutions, expand blood vessels. Opiates and narcotic substances at high doses also expand vessels. Chloroform and ether, at high dilutions, narrow the vessels. Moreover, with consistent dilutions, one type of effect was substituted by the opposite, and vice versa. So Kravkov observed the same effect, which 60 years later was reported by Benveniste. The most important conclusion by Kravkov was that "the poison as a vasoconstrictor and vasodilator, appears stronger and stronger as it is more diluted..." "... Poison dilutions at which they are still active in our experiments was 1032 but, apparently, this concentration is still not the limit for the poisons' action."

In his experiments, Kravkov also used another biological model—winter frogs possessing dark skin of a color very sensitive to various influences. He placed the frogs in the water into which were diluted the above-mentioned biologically active substances (in dilutions of up to 1024). He observed, that it took a few hours before the skin of most frogs began to fade, and the pigmentation became concentrated into crisp "islands" on the skin's surface. The usual dose dependence was not observed, but often, the effect was "more sharp at 1024 concentration than, for example, at 1023 concentration." Similar results were obtained on the isolated frog skin.

Kravkov understood that to speak of the direct effect of the molecule "poisons" on the hypothetical molecular targets of protoplasm was not possible at such dilutions. He made a theoretical calculation. At such dilutions, one molecule was dissolved in a few liters of water, or even tens of liters.

But as biological effects were clearly present, he suggested that in the process of dilution, it was as if the "...poisonous molecule melted and induced some special properties to the solution..." Kravkov expressed a bold hypothesis: "You have to think that this change is due to the collapse of the properties of the venom molecules into positively and negatively charged ions, and maybe release of the electrons from the substances' atoms."

Thus, there is a gradual transformation of poison matter into electrical energy... Under such conditions, poisons can be thought of as becoming special stimulants of protoplasm, causing it to vibrate in one direction or another. "He, however, agreed: 'If I keep saying that this activity is provided by electrical energy, I am only doing this because there are no other names for this energy. Anyway, I have no doubt that the action of substances in minimal doses and concentrations are not of a material nature and that living protoplasm is infinitely sensitive to the incessant transformation of matter into energy, and that life is closely linked, sensitive to the incessant transformation of matter into energy, and that life is closely linked with the global transformation of matter. This is the foundation of the entire protoplasm of life and its various manifestations."

In other words, the pharmacologist Kravkov was inclined to believe that the basis of the biological action of the chemicals could lie, not in "sticking" of molecules with the receptor of biological target, but in the resonance generated by the interaction of the electromagnetic fields of solution with the target.

This bold idea was based on the results of Kravkov 's other experiments regarding the transfer of biological information over electronic networks. They appear to be almost as bold as Benveniste's experiments, conducted nearly a century later.

Kravkov's experiments were based upon the works of the prominent biologist of the XIXth century, Carl Wilhelm von Nägeli (1817-1897) who discovered the so-called "oligodynamic" effect of metals—the antimicrobial effect of water "tinctures" of metals, in which detecting metal traces (gold, silver, copper) was not possible using analytical methods. Kravkov confirmed that the "infusion" of metals—even so inert as platinum, rhodium and gold (hardly soluble in water)—affect both the rabbit ear vascular tone and the frog skin. Based on the idea that the action of "tinctures" was not due to metal as a material substance, but "because of the transformation of matter into electrical energy," Krakov began experiments on the action of metals on living matter at a distance.

A copper plate was installed at the distance of 1-2 cm (0.3-1 inch) from the ear. Under the influence of copper, the vessels narrowed but after removal of the metal plate, the vessel expanded to the original tone. Kravkov's work was reported in 1921 at a meeting of Russian Physico-Chemical Society. "We may assume," stated Kravkov, "the existence of actual impact of metals and energy transfer through the air." In 1922, his work was reviewed at the Chemical Mendeleev Congress—the most authoritative Russian scientific forums. In 1923, an article by N.P. Kravkov was published in the German journal *"Zeitschrift für die Gesamte Experimentale Medizin"* (34, pp. 279-306).

In 1924, he published another article in the journal *"Advances of Experimental Biology"* (Vol. 3B N 3-4) under the title *"On the limits of sensitivity of the living protoplasm."* Unfortunately, Kravkov was unable to develop this line of his research further—he died in 1924. But his contemporaries appreciated his work. In 1926, N.P.

Kravkov was honored posthumously with the highest Soviet award—the Lenin prize, for a series of scientific contributions and works, among which was the article *"On the limits of sensitivity of the living protoplasm."*

Despite N.P. Kravkov's authority and despite the fact that the scientific community at the time officially recognized his discoveries, Kravkov's article *"On the limits of sensitivity of the living protoplasm"* disappeared from his scientific heritage and legacy. For example, in the article *"Nikolai Kravkov —a life dedicated to pharmacology"* (published in 2014 in the academic journal *"Experimental and Clinical Pharmacology"* for the anniversary of the "one of the greatest creators of domestic pharmacology," his last discoveries was never even mentioned.

Opponents of homeopathy, rejecting any arguments supporting homeopathy, will not mention any favorable articles from a prominent pharmacologist. This information is thanks to S. A. Vikulov, a practicing homeopath who uncovered Krakov's article and shares it freely on the Internet. (http://www.homeorealhelp.ru/ob_smd3.html).

Quite apart from the Kravkov's and Benveniste's works, in the mid-1980s a comprehensive study of the influence of ultra-low doses has been initiated (RMA) using various drugs on biological objects. A 30-year study has been conducted at The Institute of Biochemical Physics within the Russian Academy of Sciences, under the supervision of the Deputy Director of the Institute, Dr. Elena Borisovna Burlakova (1934–2016) who honorably served as Professor laureate of State Prizes of the USSR and Russian Federation. The results of those long-term studies revealed a number of effects, which indicate water has a key role in the information transfer process.

The first article published in the Russian academic journal *"Biophysics"* in 1986 reported that a biologically active substance (an antioxidant) has the maximum effect on the electrical activity of isolated neurons of a snail at a dilution of up to 10-15 M. Interestingly, this article happened two years before the publication of the Benveniste paper in *Nature*. Upon further investigation, they used a wide range of influencing factors: anti-tumor and anti-metastatic agents, radioprotective agents, inhibitors, plant growth stimulants, neurotropic drugs of different classes of hormones, adaptogens, immunomodulators, detoxifiers, antioxidants, as well as physical factors—ionizing radiation and non-ionizing radiation.

The levels of biological organization which manifest the action of ultra-low doses of active ingredients is very diverse—from the isolated enzymes, organelles, cultured cells, organs and tissues to animals and plants, and even populations. This does not mean that the effect may be observed with ultra-low doses of any biologically active substance on any biological object. The results of these studies have led researchers to believe that this means the discovery of a fundamentally new law of interaction of biological objects with ultra-low doses of biologically active substances. Each of these substances may correspond to a specific target, have its own amplification mechanism, and express characteristic features of metabolism.

However, at very low doses, a number of general laws can be observed. Among these are the following:

• Non-monotonic, poly-modal dose-effect dependence. In most cases, the maxima of the activity were observed in the specified dose intervals of doses, separated by so-called "dead zone";

• Change in sensitivity, usually an increase, of the biological object to the action of various agents of an endogenous or exogenous nature;

• Kinetic paradoxes, namely, the ability to capture the effect of super-low doses of biologically active substances when cell or body has the same substance at doses several orders of magnitude higher;

• Action of substance on a biological target have specific receptors to this substance in doses on the order of magnitude lower than the dissociation constant of the ligand receptor complex, which contradicts the standard model of a "lock and key";

• Dependence of the sign of the effect from the initial characteristics of the object;

• If some substance in the "normal" dose in addition to the basic biological effect has negative side effects, then while reducing doses, specific activity would be maintained, but the side effects disappear;

• Speed of the physiological response does not depend upon the degree of dilution of the acting agent;

• Enhancement of the effect of physical factors (fields, radiation) with a decrease of their intensity at certain power intervals and doses.

It should be noted that the basic laws of the biological effect of ultra-low doses findings by Elena Burlakova and Jacques Benveniste, almost completely coincide. Moreover, it was practically the same as discovered by Kravkov decades earlier.

In their research, Russian scientists, like their French colleagues, came to the conclusion that many of the paradoxes regarding the action of the ultra-low doses, (already described), cannot be explained without also recognizing the fact that water, in this situation, should acquire special properties. Take, for example, the fact that the sign and direction of the effect depends in a number of cases on the properties of the initial bio-objects. If the enzyme has high activity—under the action on biologically active substances in ultra-low dose—activity is reduced. However, if the activity is low—it rises.

Yet, the most striking thing is the extent, to which it changes. This is easily explained by the fact that in the solution of biologically active substances, the structure of water alters the structure of protein in the same way. Also, it ceases to be a paradox, only an effect of the influence on the bio-target substances when their concentration is many orders of magnitude below the dissociation constants of the ligand (receptor complex or concentrations of protein). A substance in ultra-low dose changes the structure of water, which affects the properties of the receptor.

What exactly is behind the coin "special properties of water" is not clear. So often, it is said that biologically active water has a "special structure" (which is more scientific-sounding). However, the term "structure of water" has gained favor most recently. It's filled with more concrete content, not only based on the surprising results obtained in the study of water systems with new methods, but also a new theoretical framework that could serve as a basis for the interpretation of these results. All this will now be addressed.

# What is the structure of water?

*Only with the death of dogma the truth is born.*

Galileo Galilei

The nature of the phenomena detected by Benveniste is: that as a result of a particular treatment of water, it may acquire the characteristic of a biologically active substance even in the absence of active molecules in water. This implies that water itself can be in different stable states, which are responsible for different biological activities. Since the biological activity of chemical compounds is determined by its chemical structure ("structure defines a function"), then, based on that logic, various formulations of water having different activity must somehow differ in structure. When speaking about the "s" structure," we usually imply some static situation having specific structural features.

It is difficult to imagine that the simple water molecule, in which two hydrogen atoms are attached to an oxygen atom, can exist in multiple conformations. How typical of complex chemical compounds. But, water ordered structures could only be presented in the form of ice, of which more than a dozen different crystal structures are known.

Science believes that the stable ice crystal structures exist because of hydrogen bonds between water molecules. However, they are very unstable (the lifetime of one hydrogen bond is in the range of 10-11 - 10-12 seconds), while the ice crystals are stable because each water molecule within the ice crystal is surrounded by four other water molecules, and forming with each of them, a hydrogen bond. At low temperatures even if this bond is broken, molecules do not have time to scatter in different directions, and the relationship between them is reduced.

At higher temperatures, when water is liquid the velocity—of the thermal motion of water molecules dramatically increases the probability that molecule after breaking a hydrogen bond with a neighbor, will remain in the same place—is significantly reduced. However, even in liquid form, water's many molecules may form not one but multiple hydrogen bonds with neighbors, remaining longer associated.

And even if a few hydrogen bonds break at the same time and water molecules "fall" off the associate into an environment filled with randomly moving water molecules, the defect can be restored by another molecule, and an associate will maintain its existence. Based on this approach, the idea of the presence in liquid water so named "flashing" clusters floating in a sea of non-associated water molecules was developed for decades. From the experimental data, for example, slow neutron diffraction, it follows that such ordered clusters may extend to a distance of, at least 10 Angstroms.

From many mathematical models of structures in water, the best known is a model of the mixture of two states by Frank Young and Vienna, proposed in 1957. It is, of course, out of date, but until now, it was included in most textbooks on physical chemistry. According to this model, at room temperature the formation of "blinking" clusters involves up to 2/3 of the total number of water molecules and the remaining molecules are in the form of monomers.

According to IR spectroscopy, the average number of molecules in one cluster, at room temperature, is about 90-100. Increasing the temperature will cause the thermal motion to break the hydrogen bonds, leading to a reduction in size of the clusters, and the proportion of molecules belonging to their composition.

In addition to the model of "flashing" clusters, more than a dozen different models of liquid water have been advanced in recent years. In all these models, liquid water is seen—not as a homogeneous but as a heterogeneous system—where more or less ordered clusters "float" in something like a dense gas of less-ordered water molecules. In all these models, a cluster includes from tens to several hundreds of water molecules, (i.e. a cluster size does not exceed a few nanometers). Large structures cannot occur due to the shortness of the lifetime of the hydrogen bonds, the challenge of "gluing" together water molecules and thermal fluctuations. These models, obtained by different authors on the basis of computer modeling, also differ from each other by proposing different geometric shapes of clusters and the time of their life.

Martin Chaplin, who created the most complete and continuously updated online encyclopedia of water, devoted quite a lot of space to these models and offered a beautiful encyclopedia of water. The beautiful model of "icosahedral clusters" which he offered, consists of 280 water molecules and, according to computer simulations, can all occur in water under certain conditions[15]. According to Chaplin, the presence of such clusters in water explains some of its anomalous properties (but unfortunately, not all, nor even the most important—which is to explain the role of water in biology).

In Russia, the model of Moscow biophysicist Stanislav Zenin gained fame[16]. He defended his doctoral dissertation on the structure of water. The basis of this model is the hierarchy of three-dimensional structures, based on a crystal-like "quantum of water" consisting of 57 molecules. This structure, according to estimates, is energetically favorable and may be destroyed with the release of the free water molecules only at high concentrations of alcohol and similar solvents. Due to their ability to form hydrogen bonds with each other, Zenin assumed that from "quanta water," may form various combinations having different biological activities.

This approach is typical for the kind of models that include several tens of the cluster, a maximum of several hundreds of water molecules. Crystal-like patterns, being visible and intuitive, are very convenient and lend themselves well to computer simulation.

However, water molecule structures exist due to the hydrogen bonds whose lifetime is of the order of millionths of a second. So, even if due to the "reasonable" network of hydrogen bonds, the life of such constructions will exceed the lifetime of an individual bond in hundreds of thousands or millions times. It does not provide conditions for so called, "memory of water," which lasts for hours, days and weeks. Therefore, from the point of view of conventional scientists, all discussions about the memory of water are, indeed, pseudoscience.

On the other hand, the term "structure" is applicable not only to static objects. The notion of structure may be attributed to the process localized in space and having specific geometric shape. Since this process is able to transform and move in space, it has the

---

[15] http://www1.lsbu.ac.uk/water/cluster_history.html
M. F. Chaplin, A proposal for the structuring of water, Biophys. Chem. 83 (2000) 211-221.
[16] Zenin S.V. Orientation effects in water systems. J of Phys Chemistry. 1994. 68. 500-503

characteristics associated with time, i.e. it is a temporary structure.

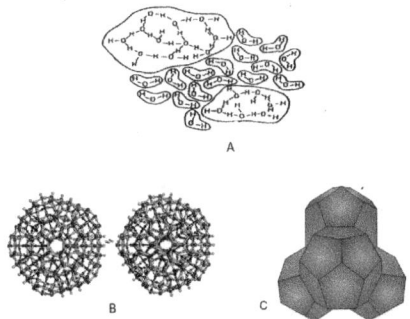

*Fig. 4. Various models of structuring of water. A – model of "flashing clusters" by Frank Young and Vienna, B - icosahedral clusters by Martin Chaplin, C - the model of "quantum of water" by S. Zenin.*

Temporary structures can be characterized by frequencies (frequency range), amplitudes and coordination (phases). It is essential that this process may exist only on the condition of a constant "flow" of matter and energy through this formation. The gaseous water forms, such as tornadoes and hurricanes, are dynamic structures, which may exceed the stability of the strongest static structures. After all, the destructive power of tornadoes and hurricanes are all well known. Ordinary clouds are dynamic structures as well.

They occur in clear air due to the increasing concentration of gaseous water in certain areas (its condensation), and may further exist for a relatively long time, often retaining its shape before it disappears. Note that meteorologists classify clouds based on their morphological features. Even with the seemingly stable form of clouds, it is hard to imagine the substance from which it is built— gaseous water —not communicating constantly with moisture from the surrounding environment. Dynamic structures in liquid water are characteristically funnels and eddies, which under certain conditions, can exist for a very long time. Thanks to observations from space, giant whirlpools with a diameter of up to tens and hundreds of kilometers were discovered recently in the oceans. The lifetime of such water bodies are months and years. From the viewpoint of fluid dynamics, based on the concepts of classical mechanics, it is very difficult to explain the mechanism of their occurrence and the support of their sustainability.

Tornado

Even ice possesses its own dynamics. Glaciers are constantly moving, creeping towards mountain valleys or crashing into the sea. So, Bear Glacier in the Pamir Mountains periodically moves more rapidly down—passing 2 km in a couple of months. On the valley side, it blocks the lake and its breakthrough leads to catastrophic flooding in Vanch Valley. It's promising to know that after many years of observation, glaciologists have learned to predict these phenomena and can confidently recommend that certain precautions be taken.

Thus, both liquid and gaseous water must be analyzed in terms of their ability to form a dynamic, as opposed to hypothetical, static structure. Dynamic structures arise, keep their shape for some time, and then disappear by flowing into a different form, a different structure. Dynamic defines the main quality of the water in any state–gaseous, liquid, and even solid. If you think about it, objects in the universe are essentially dynamic structures—from atoms to planets, stars, planetary systems, to galaxies (many of which, by the way, have the same structure as atmospheric vortices). All of these arise out of something. All of these exist with a continuous exchange of matter and energy with the environment. And all of these, sooner or later, disappear, to be transformed into other objects. But of all the diversity of natural phenomena mentioned, here we are mostly interested with water systems, and among them in living systems, which includes you and me.

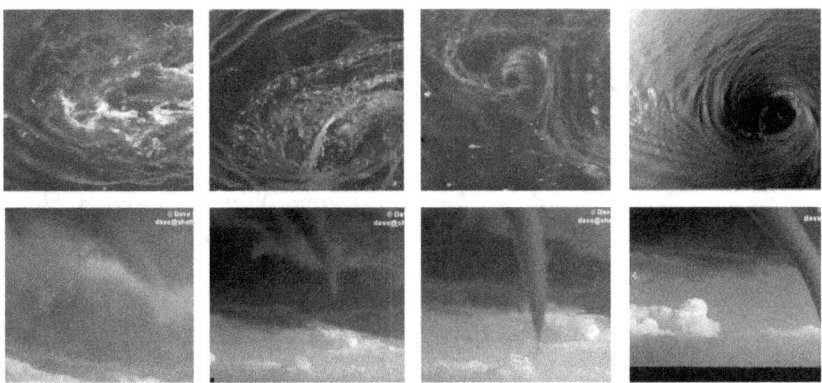

*Successive stages in the self-organization of the funnel and open water tornado*

All living organisms—from bacteria to the biosphere—are a dynamic structure of water, and water is a substance whose motion through these structures provides for their living conditions. This is illustrated by a mass of obvious examples. Amoeboid movement of many cells is the moving of gelatinous protoplasm, for 99% consisting of water, from the back of this structure to the front, and then back again. In most plant cells, cyclic movement of protoplasm was found (the exact mechanism of this movement has not yet been established). There is evidence that blood moves through the vessels, not so much due to the pumping function of the heart, but because of its self-motion[17]. These amazing properties of water systems will be discussed below. In plants, the biological fluid is continuously moving along the conducting vessels, both from the roots to the top, and from the leaf in both directions (up and down). What forces provide a fluid motion to tens of meters against gravity? Until now, a convincing answer to this question did not exist. The above facts are well known. But the mechanisms for the directional movement of biological fluids, are, to put it mildly, not fully understood. And it is clear, that without knowledge of the laws that determine the origin, development and destruction of dynamic structures in water, we fail to understand what life is.

---

[17] Marinelli R., Fuerst B., van der Zee H., McGinn A., Marinelli W. The heart is not a pump: A refutation of the pressure propulsion premise of heart function. Frontier Perspectives. Volume 5, #1. Fall-Winter 1995

# "Living water" of Viktor Schauberger

*We must look into unknown dimensions, into Nature, into that incalculable and imponderable life, whose carrier and mediator, the blood of the Earth that accompanies us steadfastly from the cradle to the grave, is water.*

Viktor Schauberger

The depth and scope of scientific knowledge about the dynamic properties of water structures is simpler than the living water systems and very limited. Decades ago, some outstanding researchers realized what amazing opportunities could be associated with dynamic water bodies and thought of ambitious prospects for the practical use of these opportunities.

One of them was Viktor Schauberger (1885-1958), Austrian philosopher and inventor, who developed the idea that there's a special energy coming from moving water[18]. Working as a gamekeeper within a logging company, Schauberger spent a lot of time observing the water in nature. From his father, he learned that in sunny days, water becomes tired and lazy, while at night and especially in the moonlight, it becomes fresh and lively.

Freely flowing water tends to create a meandering channel and flows through them, forming funnels and whirlpools. Streams or jets of water do not flow by gravity or by the shortest distance in laminar way. At first glance, that would seem to be the most plausible. But on the contrary, the water squirms and throbs. Schauberger realized that spontaneously flowing water is structured. By increasing the flow rate of the liquid or gas, it transits from a laminar to a turbulent flow. Turbulence is often associated with the concept of randomness. However, according to

---

[18] Alexandersson O. Living Water: Victor Schauberger and the Secrets of Natural Energy. – Houston, TX: Newleaf, 1982, 1990.
www.kramola.info/vesti/neobyknovennoe/viktor-shauberger-razgadavshij-tajnu-vody

modern concepts, turbulence is not chaotic, but a highly organized, orderly flow. In turbulent flows, there are spontaneously generated fractal non-linear (i.e. self-similar at different scales) waves, swirls and funnels (see. Hokusai pictures). The spontaneous formation of structures implies that shaping is an internal process, and therefore void of external factors. For example, the formation of craters and vortices occur due to the intrinsic properties of the system, and not under any influence of an external shaping factor (such as a glass funnel while stirring the water with a spoon therein). Rapid movement is one of the vital conditions required for the spontaneous formation of dynamic structures. Also integral are the conditions under which it gets rid of excess heat energy—the energy of chaotic motion. Certain conditions, such as coolness and shade and, for some reason, the moonlight, facilitate the emergence of dynamic structures. Therefore, during dynamic structuring, water cools down.

And this water, according to Schauberger, acquires special energetic properties and effects may occur that in other conditions require the application of significant external forces. One of these phenomena is movement in the jets of water of heavy logs—often having low buoyancy. Based on their performances, Schauberger designed and installed water troughs with spiral notches to facilitate the floating of timber. In one case, the rafting tray stretched for 50 kilometers and its repeated bends of the stream. The experts—hydrologists—ridiculed his design, claiming it was contrary to "common sense" because, apparently they thought, the logs will immediately blocked the flow in this tray. However Schauberger cleverly chose a specific time, when water was very cold, to send down the gutter-felled trees. For one night, a huge amount of driftwood was released into the valley. This success attracted wide attention for him and Schauberger was appointed imperial advisor on alloyed devices with a salary twice as much than the graduates.

Alas, this piece of luck did not add to Schauberger's authority among scientists. None of them were seriously interested in the ideas of a self-taught man albeit talented. Schauberger simply continued his observations of the behavior of water in nature. Many who had looked into the properties of water before him could not perceive it as unusual. But not Schauberger. He deeply pondered the phenomenon that is water. Consider the well-known fact that fish spawn upstream. To do so, they need to overcome fast-flowing rivers and cover distances of hundreds of kilometers. Schauberger wondered where they got the energy to overcome the forces of the counter-flow, especially if they stopped eating. Schauberger observed that trout and salmon at first stand still in the fast-flowing water barely moving their fins. Then suddenly, the fish would rise up against the rapid flow of the waterfall.

Moreover, during lunar cold nights, in a mountain stream, he witnessed large stones rise from the bottom to the surface and float there. However, the floaters were not just any stones and rocks, but only those of an ovoid shape. All this convinced Schauberger that the swift flow of water, when it sheds its heat energy, has special properties. What extraordinary possibilities this provides for phenomena to occur when the right conditions are met! When Schauberger poured a bucket of hot water into the creek upstream (which practically does not change the temperature of water in the creek), the trout downstream began feeling the pressure of the flowing water, tried to resist, but often could not hold and was pushed further downstream.

All phenomena, at first glance, are contrary to the accepted laws of classical mechanics and thermodynamics. Moving water is cooling, although, because of the friction with the hard surfaces, it would have to be heated. Cooling water should have lost energy. We are accustomed to the fact that the higher the temperature of the body, the greater the amount of energy is in it. But Schauberger realized that while cooling running water increases its directional (i.e. ordered) energy, ordered energy is much easier to convert into work, and with much less loss than disordered energy. This is evident from watching the movement of this water in different bodies—from fish to stones.

Schauberger put forth the idea that fast flowing water creates dynamic structures, mainly vortices (funnels) where the phenomenon of "implosion" occur. Suction forces act in the direction from the vortex periphery (funnel) to its center on one side, and on the other side moves in the opposite direction to the movement of water, thus creating a funnel. The implosion is the sum of the centripetal forces, (which before Schauberger, was never considered to be an independent force), plus the lift when a vortex is oriented vertically. Schauberger contrasted the "implosion" to "explosion," or the release of energy in the expansion of the collapsing matter. Explosion is the main method of producing energy, due to burning fossil fuels and the nuclear reactions from splitting atoms. For Schauberger, obtaining energy during explosions (expansion, destruction) for useful work (construction, movement, transformation) is extremely inefficient (indeed, the very notion of "efficiency" arose out of the use of explosions). On the contrary, as soon as the conditions for the formation of structures (such as vortices) are created, it begins to grow itself, involving a mass of water. Whirlwinds, spontaneously structured, acquire a spiral shape.

There is little internal friction between molecules forming a vortex. Its elements don't push each other as within the whirlwind; they are rotating with similar velocities and close frequencies, ie, coherently (in phase). This means that the vortex can be characterized as a coherent system in the narrowest physical sense of the term. But when it comes to the word "coherence," we consider it in its broadest sense. Coherence implies connectivity, integrity, and dynamic stability. These attributes are due to the non-local connections between the elements and long-range interactions, which penetrate the system as a whole. It is interesting to note that the formation of a vortex in liquid water occurs above a critical temperature (4°C), the exact measurement at which water has maximum density. According to Schauberger, at the same time, reducing the internal friction of a vortex, will maximize the fluidity of water. The idea seems paradoxical at first.

From the standpoint of classical mechanics, the vortex has quite unusual properties. It's useful to regard vortices as similar to self-organizing machines. As soon as favorable conditions appear, a whirlwind becomes a "machine" that builds itself. All the elements of a self-organizing "machine" are its matter and energy. The material for the construction and maintenance of its capacity, its "working fluids," is derived from the environment. Thus, the entropy of the vortex, as a system, during its formation is reduced due to the spontaneous ordering of moving substances. At the same time, its free energy (energy that can make some work) is not just growing, but is concentrated to very high potentials. So, inside a vortex in the air (tornadoes) and the vortex formed in water, electrical discharges can be observed. This indicates the ionization of matter and charge separation—and marks the appearance of the plasma state of matter.

Tornados in the atmosphere are a great example to illustrate the properties and behavior of the vortices. How striking it is, that despite the regular appearance of these important and terrible phenomena of nature, modern academic science still has not offered an acceptable explanation as to the mechanism behind their functioning. Schauberger, decades ago, realized that all the vortices—regardless of whether they are in the air or in the water (do not forget that air includes a lot of water in gaseous form)—are prime examples of self-organization, upon which spontaneous concentrations of matter and energy occur.

Based on natural principles with regards to the generation, concentration and conversion of energy, Schauberger engaged in the development of some technical devices. Unfortunately, for a number of both objective and subjective reasons, none of his implosion devices survived. However, fragmentary information is available in literature and can be found. There are rumors, which suggest that Schauberger worked on a "flying saucer" prototype—a device that uses a self-sustaining vortex to create lift. Another of Schauberger's developments is the so-called "home energy generator"—a device that was supposed to produce energy with the formal value of more than 100% efficiency through the use of "Schauberger tubes."

The "Schauberger tube" is a tube with a special and substantial spiral profile, twisting water flow into a spiral. With specific ratios and the flow rate parameters of the tube, frictional resistance was not just zero (water as it becomes a super fluid), only negative. The latter means that the self-acceleration of the movement of water takes place without the application of any external force. These striking phenomena have been confirmed in a series of special experiments by hydrologist Professor Franz Poppel in 1952 at the Stuttgart Technology University.

Within the Schauberger tubes, some chemical changes occurred in the process of the helical movement of water. These processes were associated with changes in both the properties of water and the properties of the materials from which the tubes were manufactured. Unfortunately, the details of these processes were not studied. Schauberger simply assumed that under these conditions, even the transmutation of elements can be carried out.

*Victor Schauberger with his device.*

Later on, all these decades of research nearly dropped out of sight completely. This was due to a number of reasons, one of which was the inability to explain all this phenomena in the terms of classical physics. However, in the last 10-15 years, with the advent of the new millennium, Schauberger's concepts have been revisited and began to attract new attention. His work has garnered attention from those physicists and engineers who deal with the problem of "free energy" (energy from "Physical Vacuum" and "zero point" energy). Further discovery of the concepts of Victor Schauberger can significantly accelerate developments in these areas, which will contribute to the emergence of Nature-friendly technologies.

Schauberger's discoveries are closely linked with specific, systemic properties of water as manifested at the macroscopic level. Today, it is becoming increasingly clear that the spiritual principle of "as above, so below" applies to the system properties of processes at different levels of the organization—from the nanoscopic to supra macroscopic (Universe). In the next section, we discuss water systems from the standpoint of quantum field theory, the most fundamental theory of modern physics.

# Quantum Breakthrough

> *Water is life's matter and matrix, mother*
> *and medium. There is no life without water.*
>                    Albert Szent-Dierdy

The previous section emphasized how the theoretical notions about the structure of liquid water currently dominating science, are founded on the fact that water molecules are bonded by weak electrostatic hydrogen bonds. Based on the limitations of these ideas, it is impossible to explain a number of phenomena related to the behavior of water. We have only considered two extreme examples of such phenomena—the "Benveniste effect," which appears at the molecular and supra-molecular levels and the "Schauberger effect," the self-organization of water systems at the macroscopic level.

Next, we discuss other phenomena that are likewise unable to be explained within the present framework of classical physics. Since these phenomena have been confirmed by many observations, in both controlled scientific experiments and in everyday life, a scientific explanation for these facts is wanting. What is the alternative but to relegate this information to the field of miracles and mystics?

A fundamentally new approach to the theory of water systems was developed by a group of Italian theorists who were led by prominent physicists Giuliano Preparata, (1942-2000) and Emilio Del Giudice (1940-2014). Both were specialists in high-energy physics—particularly, in quantum field physics. Important provisions of quantum field theory state that, rather than isolated particles, a set of quantum fields represent reality.

Not only are photons quanta of the corresponding (electromagnetic) field, but electrons, protons, neutrons, etc. also represent a fundamental institution of appropriate fields. But when considering the "low-energy reality" field, ideas shift into the background and the focus shifts towards the particles, which possess, among other features, paradoxical behavior. In some circumstances, they behave as fixed particles, while in others—they behave as a space wave.

Giuliano Preparata and Emilio Del Giudice

These paradoxes are an attribute of classical quantum mechanics, one of the major achievements of science. It explains the periodic table's principle of construction. What quantum mechanics does not do is allow the calculation of the configuration of complex molecules, nor does it explain the behavior of gases and liquids. For a long time, the principles of quantum mechanics were considered pertinent only on the levels of the microscopic and submicroscopic, while the transition to the macroscopic level was examined primarily using classical physical theories.

Giuliano Preparata and Emilio Del Giudice were among the first to show the many paradoxes in the characteristics of the condensed phase of liquids and solids, including water, which is, by definition, macroscopic. Explanations come from the principles of quantum electrodynamics (QED). Quantum electrodynamics is an integral part of quantum field theory, which originated in the early twentieth century. Only with the work of Schwinger, Feynman, Dyson and Tomonaga, in 1948, did QED gain recognition. In 1965, they received the Nobel Prize for their work, making this theory a working tool for physicists.

According to quantum field theory, the vacuum is not an empty space and any individual field, (an electromagnetic field, in particular), cannot disappear. The electromagnetic field (and other fields), penetrating the space ("physical vacuum") always fluctuates. Quantum fluctuations of the electromagnetic field provides the interaction of particles—protons, electrons—to generate atoms and molecules, which are themselves "clumps" of the respective electromagnetic fields and are also subject to continuous fluctuations. Thus, the main provision of quantum electrodynamics is that the interaction of particles in nature provides for the exchange of photons (electromagnetic field) fluctuating between charged particles and between them and the electromagnetic field of the physical vacuum.

The QED theory developed by Feynman, Schwinger and Tomonaga was applicable only to gases, allowing them to solve a variety of problems that have never been explained in terms of classical electrodynamics. Italian scientists extended the theory of quantum electrodynamics (on the condensed liquid phase) and obtained important results. They showed that the interaction between the electromagnetic field of the physical vacuum and matter in the liquid state leads to the appearance of the areas in which water molecules begin to oscillate in unison (becoming phased), i.e., they enter a coherent state. These "special areas in liquid" are similar to the working body of laser. The authors gave these special areas the name, "coherent domains" (CD). However, in contrast, the technical laser substance in such domains become excited, not so much by the forced pumping of energy from an external source, as by internal reasons[19].

The most interesting consequences of the Preparata and Del Guidice theory were obtained in the analysis of water CD. Their approach has allowed the understanding of why water acts as the main organizing force of nature. They provided theoretical framework to explain not only the set of "abnormal" physical and chemical properties of water, manifested in the inert (non-living) nature, but also solve the mystery of life, based on the fact that water is the basic substance of living matter, and all processes of life are based on water's physical and chemical transformations.

Thus, a cloud of quasi-free electrons surrounds the coherent

---

[19] Del Giudice E, Preparata G and Vitiello G. Phys. Rev. Lett. 61 1085–1088, 1988

domain. A very small portion of energy—less than 0.2 eV, excites the electron plasma. Excitation turns the electron plasma into a micro-vortex rotating around the coherent domain. Thus, any external noise can convert coherent domains of water into vortices, coherent vortices. For example, when water is stirred or shaken, small amounts of free energy are capable of producing the rotation of the vortices belonging to the coherent domains when the quasi-free electrons begin to rotate at kilohertz frequency. This rotation is motion without friction. The coherence implies that the whole mass of molecules is moving more like a ballet without collisions than a crowd. Motion without friction can last a very long time, which distinguishes it from the lifetime of the excited state of a single atom, which lasts, somewhat about 10-10—10-11 of a second. Motion without friction may not subside for many hours, weeks and even years.

This property of coherent domains may explain homeopathic medicines. In an individual atom, the excited state is so short in durance that after the first excitement, there is an almost instantaneous relaxation into the fundamental state. Then, the second excitation merely repeats the cycle. Therefore, the system of individual atoms cannot accumulate excitation.

In contrast, when you shake a drug containing coherent domains, in order to make homeopathic medicine, you must provide an impetus for an excited rotational state of quasi-free electron of coherent domains. This stimulation can live a very long time, and does not subside in a split second. So, you have stirred up the water and achieved the rotational excitation of, say, 30 kHz. You stirred up the water again, causing a new excitement, which gets added to the previous one. A remarkable feature of the coherent domains is that, thanks to a long lifetime of the excited states, excitation may increase. For example, the first excitation gives 30 kHz, the second is 30 kHz, altogether 60 kHz, the third is 30 kHz, etc. If the lifetime of each excited state is, say, 1 hour, and between the shaking the interval is, say, a few moments, you can produce a powerful rotational excited state because it is based on a coherent electron motion.

Here is another important point. The rotation of the electrons is equivalent to the rotation of the electric current and produces a magnetic moment putting us in a very favorable position as we live on Earth, which has its own permanent magnetic field. Furthermore, the axis of the magnetic dipole of the domain is parallel to the direction of the magnetic field of the Earth, and all the excited domains are aligned parallel to each other.

But if large bio-molecules appear in water, they become the basis for the formation of coherent domains around them. So, instead of spherical configurations, they become tubes coaxial to the axis of the molecule. Water is coordinated around its axis. In this sense, the coherent domain coincides with the bound water. Since coherence is a preferred state, and since it reduces the total energy, the components of bio-molecules are beginning to participate within the coherent mode. Let's say that the water molecules dance with each other, and amino acids of the protein join this dance. In this case, they acquire the same frequency as the coherent oscillation.

Let this process be represented in a dance club metaphor. The patrons who are sitting at tables and milling around represent the disordered component of water. The people who continue to move together using rhythmic dance motions represent water's orderly component. An increase in water temperature in this instance, would equate to putting the club's patrons in good spirits by increasing the rhythm of the music. We notice more people leaving the tables to join the dancers. However, at the same time, just as many exhausted club dancers grab the tables' newly vacated seats.

We can see, that even at the same "temperature," both dancing and sitting people are constantly exchanging places. Some sit down to rest, while others get up to dance. The overall ratio of people dancing to people sitting always remains the same. In particular, this explains, the nonlinear dependence of the density of water on temperature. Ordered clusters of molecules have a lower density than disordered. Ordered clusters vary little with temperature. This can directly be correlated with there remaining a constant number of tables at the dance club, regardless of the mood of the patrons or the intensity of the music.

Pierre-Auguste Renoir. *Ball at the Moulin de la Galette*

These statements are not unfounded; they have a sound scientific basis. To further convince the reader, it makes sense to give voice to Emilio Del Giudice, one of the authors of the theory of water, based on the principles of QED. As a talented teacher, he could explain incredibly difficult concepts in simple and approachable ways while maintaining the full scientific rigor of his subject. His lectures, publications, and even his communications provided tremendous intellectual and emotional pleasure and were treasured by all who were fortunate enough to know and correspond with him. In particular, Vladimir Voeikov worked for many years with Emilio Del Giudice. Together, they published a series of joint projects.

We acquaint the reader with the teaching style of Emilio Del Giudice. This example of excerpts comes from his article published together with Paola Rosa Spinetti and Alberto Tedeschi on the on-line journal *Water* 2010, 2, 566-586; doi: 10.3390 /w 2030566 (not to obstruct this text, we eliminated references, for full text readers may turn to the original article):

"Just considering the existence of the fluctuations of the quantum vacuum, as far back as in 1916, W. Nernst suggested the possibility of tuning together the fluctuations of all the components of a system and therefore the appearance of a common phase.

This possibility has been checked in the framework of QED, where the interaction among atoms mediated by the e.m.f. is addressed, starting from first principles. We will summarize here the main points of this approach, by avoiding all the difficulties of the rigorous mathematical treatments and the technicalities of QED. We will use intuitive arguments extensively to the benefit of non-specialists, paying the price of some lacking precision in places.

Our starting point is an ensemble of a large number N of atoms (or molecules); For the sake of simplicity, we assume that they have two states only—the ground state and the excited state—whose excitation energy is $E = h\nu$. This assumption will be dropped eventually. The size of an atom is in the order of 1 Å, whereas the size of the photon able to excite the atom is its wavelength $\lambda = c/v$, which, in the case of an excitation is in the range, as usual, of 10 eV, would be in the order of about 1000 Å. "Therefore, the size of the object able to induce a transition in an atom is about one thousand times larger than the atom! Just this mismatch is at the origin of the possibility of producing extended regions where component atoms are correlated. As a matter of fact, one (virtual) photon, which got out from the quantum vacuum because of Heisenberg quantum fluctuations, could excite an atom with a probability P in the order (according to the estimates based on the Lamb-shift) of $10-4 \div 10-5$.

The excited atom would decay after its typical decay time, giving back the photon, which could alternatively be reabsorbed by the vacuum or excite another atom. The relative probabilities of the two events would depend on the density $n = N/\lambda 3$ of the atoms present within the volume $\lambda 3$ of the photon. When the density n exceeds the threshold ncrit, the photon would never be able to reach again the vacuum and will bounce forever from one atom to another within the volume $\lambda 3$. Therefore, the vacuum has given a photon to matter. This process would go on until many photons get trapped and a sizeable electromagnetic field (EMF) is built in this region. This field produces two consequences:

(1) It attracts co-resonating atoms, which are, of course, the atoms of the same species, producing a large increase of density, as observed in the phase transition vapor-liquid; the saturation density corresponds to the inter atomic distance at which hard

core repulsion becomes important. So the observed density would depend on the short-range forces.

(2) It produces a common oscillation of all the trapped atoms giving rise to a common phase within the whole region, which for this reason is named Coherence Domain (CD). This common phase of oscillation does not coincide with the original phase of the free photon, since the time of oscillation of the photon involved in the common oscillation with atoms should be supplemented by the time spent within atoms in the form of excitation energy.

In this process, atoms and photons have lost their original identity, giving rise to energized matter (the *energid* of Sachs) made up of quasi-particles entangled among them in the CD. Just for this reason, photons cannot get out from the CD; their squared mass $m2 = h2 \, v2 - h2c2/ \, \lambda2$, which is zero for a free photon, becomes negative for the quasi particle (self-trapped photon) because of the above-mentioned increase of the oscillation period and hence of the decrease of the frequency. A negative mentioned increase of the oscillation period and hence of the decrease of the frequency. A negative value of the squared mass implies the impossibility of propagating and consequently the quasi-particle would remain trapped within the CD. This is quite a fortunate result, since it guarantees the stability of the system, which would otherwise continuously lose energy.

The concentration of energy in a small number of microstates (in principle, just one) from the original large number of microstates (corresponding to the many configurations of the uncorrelated atoms) implies a large curtailing of entropy. Unless a corresponding amount of energy is released outwards, this would violate the Second Law of Thermodynamics. This release of energy gives rise to an energy gap. An energy gap implies that the energy of the coherent state is lower than the energy of the original non-coherent state. The energy gap prevents the occurrence of a Perpetuum Mobile. The energy originally lent by the quantum vacuum is given back through the above outflow of energy, which is nothing else than the latent heat of the phase transition. The conservation of energy and the Second Law of Thermodynamics are therefore satisfied.

It is apparent that the above Quantum Field theoretical predictions agree completely with the observed pattern of a vapor-liquid transition. In the conventional approach, computer simulation techniques have allowed investigation of the properties of water clusters made up of a small number of molecules and their stability. These investigations have recognized that many body forces play an important role. In other words, condensed matter cannot be described only in terms of pair potentials.

The electro-dynamic attraction, induced by the onset of coherence, is supplemented in real systems by the short-range static attractions. Static attractions could only play a role after molecules are brought into close proximity by QED attraction. We should point out that the static attraction does not occur between the molecules in their individual ground state but among molecules in their coherent state, where there is a significant contribution to the excited state. We will see that this consideration is very important in the case of water."

In the real case of molecules having not just two internal configurations but many, the choice of the pair of states involved in the coherent oscillation demands the estimate of the time required by the onset of the coherent regime. The pair of states, which eventually give rise to the coherent state, is the one having the fastest rising time toward the state where the coherent oscillation appears. In the case of water, this time is estimated to be in the order of 10–14 seconds.

Let us now discuss the peculiar case of water. The above theory applies to all molecular species. However, in the case of water, the excited state involved in the coherent oscillation (12.06 eV) lies just below the ionization threshold of the molecule (12.60 eV). An oscillation of 12.06 eV corresponds to a water CD the size of 0.1 microns. The onset of the coherent oscillation gives rise to the appearance of one quasi-free electron in the coherent state; therefore, the CD becomes a reservoir of quasi-free electrons that are easily excitable.

The spectrum of excitations of water CDs has been derived; each excitation corresponds to a coherent cold vortex of quasi-free electrons. Actually, quasi-free electrons belong to a coherent state so that an external perturbation that is smaller than the energy gap, cannot be received by any individual molecule, but is stored by the CD as a whole, giving rise to a collective excited state, which is still coherent. The existence of a huge number of excited states, characterized by their angular momentum L, whose energy spacing is in the order of radio-wave energy (some tens of kHz). Since the vortices are cold, they cannot decay thermally, thus their lifetime depends on the lifetime of the parent CD. Consequently, the excitations of CDs could last a very long time and, moreover, give rise to a sum of several subsequent excitations whose energy therefore becomes higher and higher. The possibility of the storage is increased by the coupling of the magnetic moments of the cold electron vortices with the Earth's magnetic field, which aligns them.

The spectrum of the excited states of the water CDs, is limited upwards by the energy gap which is 0.26 eV per molecule; since in a CD there are about six million molecules, it is apparent that the spectrum of an isolated water CD has practically no upper limit. This means that within the CDs it is possible to store amounts of energy that can reach the visible and the ultraviolet. In this way, the water CD would become a device able to collect the energy coming from the environment and transform it into energy able to induce electronic excitations in the bio-molecules surroundings the CDs.

This property, which emerges naturally in the scheme of QFT, implements the requirement of Szent-Gyorgyi, made long ago on purely biological grounds. This result opens new perspectives in the investigation of important natural phenomena such as lightning. A lightning bolt emerges from the clouds, which are ensembles of droplets of water suspended in air and nothing else. In spite of this simplicity, lightning carries huge amounts of energy and electric charge. Since we have proven that a water CD can easily release electrons and can store huge amounts of energy, we are faced by the appealing possibility that we could learn something about the dynamics of lightning by using the QFT approach.

We conclude this section by discussing the structure of the short-range static forces among the coherent water molecules. The excited state appearing in the coherent state of water is a 5d state, namely, in this state there is a very decentralized electron having a high angular momentum ($L = 2$).

The electron cloud in the excited state assumes, therefore, a torpedo-like shape. The average shape of the electron cloud in the coherent state is the combination of the shapes in the two component configurations between which the coherent molecule oscillates. The contamination of the electron configuration of the excited state induces the appearance in the ground state electron cloud of two protuberances oscillating with the same frequency as the collective oscillation of CD. This produces the observed phenomenon of the H-bonding, which becomes therefore an effect of the existence of the coherent regime. So the H-bonded network of water molecules is the phenomenological appearance of the coherent fraction.

If a water coherence domain were to accept a small number of guest molecules among its participants, then the excitation energy stored in the CD would become available to the guest molecules. When the amount of stored energy matches the activation energy of the guest molecules, the energy would be transferred to them, simultaneously producing their chemical activation, the energy discharge of the CD and a chemical reaction array. The CD would then behave as a multimode laser, causing a number of consequences to arise:

1. The CD has completed an oscillation whose duration depends on the rate of energy storage, on the height of the required activation energy and on the rate of chemical reaction. The inverse of this time is the frequency of oscillation of the CD. Should many neighboring CDs be in the same chemical and thermodynamic environment, they could enter into a collective coherent oscillation that would, in turn, increase the degree of coherence (which is the width of the coherent oscillation frequency) of each of the participant CDs.

2. The chemical reactions, which occur on the surfaces of CDs and can benefit also from the electron transfer available there, are no longer governed by diffusion but are governed by electro-dynamic attraction. According to a theorem of QED, two molecules oscillating with frequencies $v1$ and $v2$ within a region filled by an electromagnetic field oscillating with a frequency $v0$; develop a very strong attraction when the three frequencies coincide. This long-range attraction replaces diffusion as the molecule interaction agent. The existence of codes governing the array of biochemical reactions could therefore be understood.

3. The energy output of the chemical reactions is released because of coherence as an excitation of the electromagnetic field trapped in the CD and absorbed by the water CD. A corresponding shift of the CD frequency is produced, in turn changing the molecular species able to be attracted, consequently opening a new biochemical cycle. Each cycle is therefore opened by the outcome of the previous one. The possibility of an ordered array of

biochemical reactions emerges.

Therefore, two correlated dynamics are at work: (a) the emergence of an extended coherence among coherence domains depending on the frequency of the CD oscillation governed by the energy charge and discharge processes. (b) The emergence of a time-ordered biochemical array governed by selective attractions among molecules. The two dynamics are tightly interconnected so that one could say that biochemistry is the tool necessary to keep water organized in a long-range.

The onset of coherence among coherence domains stabilizes the coherent fraction of water since the energy gap of the additional coherence adds up to the energy gap of the single CDs, providing an additional protection against the disruptive effect of thermal noise. Consequently, liquids endowed with this extended coherence exhibit a less flickering internal landscape, giving rise to the possibility of bulk water displaying coherent patterns. This is exactly what's observed in the special waters produced recently. We also observe that coherence among coherence domains is not sufficient to adhere the CDs together, but implies simply that separated domains oscillate in unison.

Accompanying the formation of an ensemble of CDs is the expulsion of the solutes, including the atmospheric gases from their inside, so that in the very moment of CD formation in the bulk liquid, a micro-bubble should appear also. In normal bulk water, micro-bubbles appear and disappear in a flickering way— mirroring the flickering space distribution of CDs. On the contrary, in the special waters where the extended coherence is established, the space distribution of CDs flickers much less. This is mirrored by a non-flickering, ordered space distribution of micro-bubbles. Therefore, the transition between flickering and ordered arrays of micro-bubbles in liquid water reflects the transition between the coherence and non-coherence of the water CDs.

## The Dialog between Liquid Coherent Water and the Environment:
## The Emergence of Time Evolving Information

The existence of a dissipative structure made up by the coherent array of water CDs, whose extended coherence depends on the presence of non-aqueous guest molecules in water, endows this liquid with the capability of communicating with the environment.

Long ago, Giorgio Piccardi reported that significant changes in the physical properties of molecular systems suspended in liquid water occurred simultaneously with cosmic or environmental events. For instance, he detected changes in the precipitation rates of colloids following the time evolution of sunspots or climatic events. As a matter of fact, water appeared as an accurate probe for a large number of external events. This property sheds possible light on the capability of living organisms, which we know to have a dominant content of water, to perceive external events.

More surprising, water and living organisms are shown to be able to perceive very subtle events, also below the resolution threshold of technical devices. We like to quote in this context the research of V. L. Voeikov and his Russian colleagues, who were able to detect peaks in the amount of photons emitted by water added with bicarbonates and Luminol coinciding with Sun and Moon eclipses, and also earthquakes occurring very far from Moscow, where the lab was located.

Keep in mind that "pure" water does not have the property of sensitivity described here. Only water containing solutes or suspended particles. Like colloidal solutions, it is sensitive in this way. This property recalls very much the conditions necessary in the QED theoretical approach for obtaining extended coherence in water.

In the QED approach, the surprising properties found, among others, by Piccardi and Voeikov, do not appear surprising at all. Actually, water CDs contains trapped electromagnetic fields, which produce a magnetic vector potential A in the surrounding space, whose rotor—and hence the magnetic field—is zero [we recall that H = rot A]. A extends on a much longer range than H, since this last field is given by the space derivatives of A. There is, therefore, a coupling between the vector potential produced by the water CDs and the vector potential originating in the electromagnetic dynamics, which occur in the environment—like the electromagnetic radiation produced by sunspots, cosmic events, atmospheric events and movements in the terrestrial crust. According to which the phase of the system is changed by the magnetic vector potential, this situation is exactly what could give rise to the Bohm-Aharonov effect. As a consequence of this effect, coherent systems (in general and aqueous systems, including living organisms in particular) are very sensitive detectors of weak magnetic fields through the detection of their magnetic potential.

The evidence collected for many years about the impact weak magnetic fields have on aqueous and living systems—so far unexplained—could at last find a rationale in the existence of electromagnetic structures in the supra-molecular organization of liquid water. Very recently, the group led by L. Montagnier has been able to detect experimentally, the presence of electromagnetic signals originating in the water surrounding bio-molecules.

Since the onset of the extended coherence of liquid water depends on the presence of non-aqueous molecules, rocks containing some water provide a useful non-biological model system. According to the dynamics described in the previous section, the water in the rocks—through the chemical reactions among the carbonates and the atmospheric gases—produces water having higher coherence than plain bulk water. The strange phenomena observed in water come from springs located in caves. These could be analyzed just in this context. Actually, rocks and living organisms have the presence of an interfacial layer of coherent water in common, which appears like a reasonable candidate to be the actor of the reported effects.

## Conclusion and Outlook

Quantum Field Theory has produced a vision of liquid water as a medium. A peculiarity of the molecule electron spectrum reveals itself as an essential tool for long-range communications—being able to change its supra-molecular organization according to the interaction with the environment. The electromagnetic fields trapped in the coherence domains and in their coherent arrays produce electromagnetic potentials governing the phase of the whole system, which in turn gives origin to selective attractions among the solute molecules. In this way, an array of biochemical reactions (soma) and time-evolving information simultaneously evolve, leading to the appearance of the self-consistency, which opens a new perspective for self-maintaining and stability of the systems under study."

We have seen how, in recent years, the theories of Propagate and Del Del Guidice provided clear evidence. Later, we will look at some of the most interesting results in our opinion. Though we did not set the task to review all that has been done in recent years, much of this data was presented at the Annual Conferences on the Physics, Chemistry and Biology of Water, held under the auspices of Professor Gerald Pollack (www.waterconf.org) and published in their online journal www.waterjournal.org.

Emilio Del Giudice, Vladimir Voeikov and Konstantin Korotkov at St. Petersburg conference in 2012.

# Water as a Complex Multicomponent System

As mentioned in the previous section, it follows from the theoretical principles of Italian quantum physicists that liquid water should be a multiphase, non-equilibrium and, therefore, the active complex system.

Physicists and mathematicians only recently began to investigate the properties and behavior of such systems. At the same time, according to the dominating model of liquid water, the so called "flickering cluster" water model should be regarded as similar to a dense gas: low structured set of small simple $H_2O$ molecules. As there are a lot of experimental evidences supporting this model, many scientists are opposed to considering water as a complex system.

But the Preparata and Del Giudice model does not deny that significant part of water is represented by "dense gas." Their model implies that CDs and much less structured water co-exist and interact with each other. That alone makes water a complex dynamic system with much richer behavior than that of any homogenous matter. For example, if water is, indeed, an active complex system, under certain conditions it should not only change its state in response to the weak resonance signals, but also for a long time maintain such a condition. This property may be known as the "memory of water." Moreover, if water is a dynamic non-equilibrium system, by itself, it can be a source of various signals.

In the last decades, convincing experimental evidence has appeared, supporting the theoretical model of Italian physicists. It was shown, for example, that a sufficient condition for the emergence of a long-living organized phase in aqueous systems is the presence of the interface of liquid water with hydrophilic surfaces with which it contacts, and even the interfaces of water with air. The Professor of the University of Washington Gerald Pollack received this evidence. He has developed a simple experimental model, using which only in a few years had made a number of discoveries showing how complex is even very clean, but "real" water. By "real water" we mean water, which always contact with the vessel border, may contact also with the air, and in which non-aqueous substances are always present. All these conditions promote water separation into two different substances (phases). Interaction of these phases having very different properties is the cause of many phenomena, which seem mysterious to those who are considering water as simple, super-clean and limitless "liquid gas."

Before we discuss the details of the discoveries made by Pollak and his colleagues, we note the impetus for his research that never fully disappeared was the scientific understanding about the special properties of water in living matter: in particular, in the cell protoplasm. Having a long history, is the idea that water in living cells, filled with proteins, polysaccharides, and nucleic acids, must be structured due to the interaction with all these particles and so must differ in its properties from "normal" water. Since the beginning of the last century, many prominent physiologists, biochemists, chemists, and cytologists have defended this idea. They proceeded from the obvious to any unbiased scholar facts, and thus formed the basis of colloid chemistry.

One of the supporters of the concept of highly organized structure of intracellular water was an American chemist and biochemist Ross Aiken Gortner (1885-1942). He drew the attention of his colleagues to living creatures, like jellyfish. More than 99% of the jellyfish mass is water, i.e. the mass fraction of proteins and other biopolymers, salts, etc. in the body of the jellyfish is negligible. And although the water in the body of jellyfish is represented with the same molecules as any other water, "living jellyfish water" is very different in its properties from "normal" water. It contains much less salts than the seawater in which it lives, despite the absence of any impermeable for salts film on the surface of jellyfish. It is obvious that the condition of water in jellyfish is determined by the biopolymers, although they represent only a tiny fraction of the mass of the body[20].

Gortner received a lot of experimental data confirming the special properties of "bound" or "structured" water in a variety of colloidal systems, in particular, in bio-colloids. His findings were published in prestigious scientific journals, and a section on the special properties of water in living matter took a quarter of the

---

[20] Henry M. The state of water in living systems: from the liquid to the jellyfish. Cellular and molecular biology. 51, 677-702, 2005.

nearly 800-page book by R. Gartner, *"Fundamentals of biochemistry,"* which withstood several editions from 1929 to 1950.

Prominent Russian biochemist and plant physiologist Vladimir Lepeshkin (1876 – 1956) held a similar position. He showed that the protoplasm emitted by damaged plant cells, when shaken with water, is divided into many small droplets, not miscible with water. To imagine that the mixing of the liquid content of droplets with water is prevented by lipid membrane instantly formed on the surface is impossible–there are no such stocks of phospholipids in a cell from which you can build a bi-layer membrane.

So Lepeshkin suggested not miscible with the external water droplets of protoplasm constitute the supra-molecular complexes, formed by the water of the protoplasm, forming a connection with the "lipids" and proteins. Moreover, a flash of UV radiation accompanied cell death when exposed to cell-killing factors, such as high temperature and poison. Lepeshkin termed the UV flash, "necro-biotic rays." Protoplasm coagulates, in this process, separate into free water and denatured biopolymers. Based on this, Lepeshkin suggested that in living cells, the complexes of water with its connecting components are active. They exist in a "charged" condition and in their destruction; they lose energy in the form of radiation.

A significant contribution to the development of ideas about the special status of water in cells was made by an outstanding Russian cytologist, a member of the USSR Academy of Sciences, founder of the Institute of Cytology of the USSR Academy of Sciences in Leningrad: D. N. Nasonov (1895-1957) and his followers[21]. Nasonov considered the protoplasm as a colloidal phase in which the water condition is different from the conditions outside a living cell. The different states of water in the cellular and extra-cellular phases prevent these two phases from mixing (and therefore, this concept is called the phase theory of protoplasm). Water in protoplasm differs in solvent capacity from the external water and the uneven distribution of substances between cell and environment should be explained—not by the presence of specific pumps and channels in a hypothetical semi-permeable membrane that separates cell from the environment—but by different distribution coefficients between the two aqueous phases.

Nasonov drew attention to the excited non-equilibrium state of

---

[21] Nasonov D.N. Protoplasm reactions. Moscow. 1959

the protoplasm of living cells. He noted that in dark-field microscopy, living cells "in response to stimulation of the protoplasm, begin to glow with a pale blue color and then on the cell surface bright glowing white structures appear." In addition, if the material substrate of living cells is in an excited state, its optical properties should be different from the properties of the same matter in a quasi-equilibrium state.

D. N. Nasonov writes: "...under ordinary microscopy, in transmitted light, in normal cells, the nuclei is virtually invisible and can be detected with difficulty only by a light restricting boundary. The rest of the core is structure-less, "optically empty. In dark-field microscopy, under the action of any irritant in the nuclei, appear structures previously unseen, sometimes, even earlier than structures in cytoplasm. We can see nuclear skeleton and lumps of chromatin the same as we see in fixed specimens." Thus, according to Nasonov, Lepeshkin, and Gurwitsch, "living matter," the prototype of which is the cell protoplasm, is not just a water-biopolymer system with specific properties of water, but the system in the excited state capable of changing its properties in response to various physical stimuli and to perform useful work— the work to maintain life.

The most solid and logical phase protoplasm theory, developed by the US biologist Gilbert Ling (born 1919), who called it a theory of the "Association-induction," acknowledges the difference in physical-chemical properties of intracellular and extra-cellular water[22]. It also takes into account the livelihoods associated with reversible changes in the state of intracellular water, i.e., its dynamism. The theory of the "Association-induction" allows, according to Gilbert Ling, explains such manifestations of cell activity as the mechanism of energy transformation in living matter into useful work (the essence of bio-energy), muscle contraction, active transport, phenomena of growth and development, as well as cellular pathology, like carcinogenesis.

The word "Association" as part of the theory's title indicates a close interaction and interrelatedness between proteins, water,

---

[22] Ling, Gilbert Ning The Physical State Of Water In Living Cell And Model Systems. Annals of the New York Academy of Sciences. (2006). 125 (2): 401–417

and potassium ions. The idea of this collaboration came to Ling in the early 1950s when he designed his ideas in the form of "the theory of fixed charges of Ling and theory of multilayer organization of the polarized water." These theories were designed to explain why the contents of K+ in the cell is significantly higher than in the medium, while Na+ contents in the cell are much lower than K+ and often lower than in the environment.

The non-equilibrium distribution of these ions is not due to selective permeability in a hypothetical semi-permeable membrane nor is it due to the presence of hypothetical molecular pumps in the membrane. According to Ling, the presence of K+ in the cell is not in a dissolved state. Being a part of cellular proteins, it's fixed to negatively charged residues of aspartic and glutamic amino acids, whereas Na+ is not capable of it. The distinction between the ions of sodium and potassium is due to their different affinities to water: the potassium ion simply loses the hydration shell easier, which may more strongly hold the fixed negative charge over the sodium ion.

But, to have a sufficient amount of K+ ions in the cell, the protein molecules of the protoplasm must contain an excess number of negatively charged amino acid residues, and these residues should be available for interaction with K+. To do this, the protein molecule must be in an "expanded," fibrillar state—and not in the condition of the globule. Ling realized that the huge difference between the properties of fibrillar and globular proteins lies in how differently they interact with water.

*Residues of amino acids bound in protein molecule by peptide bonds:*

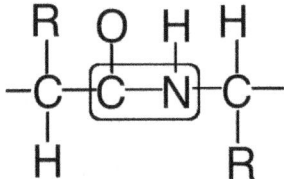

*Peptide bond (highlighted)*

Carbonyl (C=O) and amino-group (N-H) peptide bonds have, respectively, partial negative and positive charges and provide conditions for the adsorption of water molecules on the surface of the protein dipoles. Associated water molecules are polarized and additionally attract other water molecules, which are arranged around the elongated protein molecules in a multi-layer organized water structure that possesses very different properties than unorganized water. Only a small number of water-soluble proteins isolated from living matter are in the unfolded state. Most of them fold into a globule, in which the C=O and N-H groups of different parts of protein molecules cancel each other out. Such unfolded protein is gelatin–a product of partial hydrolysis of collagen, the main protein of connective tissues.

Usually located in the fibril of this protein, CO and NH groups orient water dipole molecules, which form a multilayered sheath of water along the molecule. The result is a supra-molecular structure, a jelly-like element of the system. Under certain conditions, depending on the concentration of gelatin, temperature, and salt composition, the water protein complexes merge into a single associate called "jelly" or gel. The water in the hydro-gel has very different properties from "regular" water. Even everyday experience tells us that, if we place pieces of jelly in water, these two phases can practically coexist indefinitely without mixing. Although the two, in reality, represent water, they differ in the composition of the non-aqueous components within the gel's composition.

Szent–Gyorgyi also emphasized water's special state of protoplasm. The authors of numerous publications appearing in the pre-second world war period, estimated the thickness of the interfacial water layer (near the hydrated surface) to be several hundreds and thousands of water molecular layers, instead of only one or two molecular layers, as commonly believed. Water in the state of protoplasm differs from "bulk" water in many ways. Its many physical characteristics include dielectric permittivity, freezing and boiling points, as well as a possession of liquid-crystalline properties. This implies there's a long-range order at which the molecules exhibit collective behavior. Szent-Gyorgyi observed an important implication of the liquid-crystalline state of interfacial water when its "structure" temperature decreased. This should affect the efficiency of the biochemical processes occurring in the presence of such water.

So far, the significance of interfacial water in cytology and physiology has clearly been underestimated. Simple examples could demonstrate the important role water plays in living systems as is represented by interfacial water. For example, there is one hemoglobin molecule per 7000 water molecules in an erythrocyte. A comparison of the sizes of the hemoglobin and water molecules shows that, under uniform distribution of hemoglobin molecules in erythrocyte, 2–18 water molecular layers can solely separate two protein molecules. Kuzin and Trincher came to conclude that in such thin films, water should occur in a special state that is characteristic of neither "bulk" water nor ice[23].

The water in erythrocytes possesses a quasi-crystalline structure, and the molecules form a complex spatial network whose loops contain hemoglobin molecules. Appropriate calculations showed, that in blood, nearly all the water occurs in the interfacial state. Three out of six liters of human blood is accounted for by blood plasma. At the same time, the surface area of only erythrocytes present in blood is approximately 5000 m2 (50000 square feet). Hence, even without taking into consideration other particulate elements of blood and plasma proteins, which are also hydrated, the water layer on the erythrocyte surface cannot exceed 0.6 mcm. which are also hydrated. The water layer on the erythrocyte surface cannot exceed 0.6 mcm. Even these simple calculations cast a doubt to the common opinion that intracellular and intercellular water may be looked upon as usual bulk water, as a simple solvent for water-soluble compounds.

Naturally, this raises the question of why are the two theories— the phase theory of protoplasm and the association-induction theory—both supported by a huge body of experimental data (and disproved by nobody) not more generally known? That water in protoplasm has different properties from bulk water lies in dramatic contradiction to the commonly accepted membrane permeability theory. Maybe, this is the reason. According to the theory, the semi-permeable properties of lipoprotein membranes covering a cell explain the uneven and non-equilibrium distribution of ions and other solutes between the cell and its environment. This collision was discussed in the monograph by G.H. Pollack. He came to a conclusion that the matter distribution between the cell and environment can better be explained by various versions of the phase theory than by the membrane theory.

---

[23] Kuzin A.M., Trincher K.S. Change of radio-sensitivity of erythrocytes. Biophysics. 1960. 5. No 5. C. 533–541.

# The Fourth Phase of Water

*Everything on the earth bristled, the bramble pricked and the green thread nibbled away, the petal fell, falling until the only flower was the falling itself. Water is another matter, has no direction but its own bright grace, runs through all imaginable colors, takes limpid lessons from stone, and in those functionings play out the unrealized ambitions of the foam.*

Pablo Neruda,
Chilean poet and politician (1904-1973)

Since the beginning of the new millennium, probably the most important contribution to the emerging science of water was made by Gerald Pollack, Professor of Bioengineering from the University of Washington. He gained fame in 1990s as a researcher of the mechanisms of muscle contraction. He was published in leading scientific journals such as *Nature* and *Science*, wrote several books on the mechanisms of biological motion, including *Muscles and Molecules: Uncovering the Principles of Biological Motion*, (1990), which won an "Excellence Award" from the Society for Technical Communication.

However, according to his own assertion, Pollack became less and less satisfied with the explanations of the mechanisms of muscle contraction that neglected to take into consideration the role of the major material component of living matter—water. Pollack acquainted himself with the alternative views regarding the principles of cell physiology based on the phase model of protoplasm. He discovered the material already covered here—especially resonated with the theory of Gilbert Ling and the works of D.N. Nasonov and his follower, A.S. Troshin. Pollack deeply analyzed the works of many other authors who were studying the role of water in biological functions who presented evidence concenring the special properties of water in cell protoplasm.

Working his way through the material, he became convinced that this concept could serve as an explanatory basis for most, if not all, problems of cell and organismal biology.

In 2001, G.H. Pollack published the book *Cells, Gels and the Engines of Life: A New, Unifying Approach to Cell Function* (Ebner & Sons). Immediately a bestseller, it was intensely reviewed in many scientific journals, including *Science, Nature, Cell, Immunology* and *Cell Biology, Journal of Cell Science, Trends in Microbiology*, and many others. In 2003, this book received Top Prize (Best in Show) by the Society for Technical Communication.

"The Most significant scientific discovery of this century."
*Mae-Wan Ho, Director, Institute of Science and Society, London*

THE
FOURTH PHASE OF WATER
BEYOND SOLID, LIQUID, AND VAPOR
—
BY DR. GERALD H. POLLACK

In his book, Pollack emphasizes the role of cell water and the gel-like nature of the cell in the mechanisms of communication, transport, contraction, division, and other essential cell functions. Pollack's central hypothesis is that the contents of cells are not aqueous solutions—as biochemists have tended to assume. Rather, the cytoplasm is a complex gel. Indeed, much of the behavior of cells can be explained by gel-specific concepts, such as phase transitions and exclusion of specific solutes from the gel matrix. This message was so deeply substantiated and convincing, that some of the reviewers of this book characterized it as "a 305 page preface to the future of cell biology." After publishing *"Cells, Gels and the Engines of Life,"* Gerald Pollack and his associates launched the original experimental project to examine the validity of the thesis advanced in the book.

To study the specific properties of water hydrating hydrophilic surfaces (interfacial water), Pollack developed a simple, but very fruitful experimental model. Based on the circumstantial evidence of his predecessors, that water adjacent to hydrophilic surfaces— such as surfaces of biopolymers, other hydro-gels, cell membranes, etc.—differ in their dissolving properties from the bulk water. Directly, he tried to visualize interfacial water. To determine if interfacial water excludes the solutes, he added hydrophilic or charged colloidal and molecular solutes to water, contacting large hydrophilic or charged surfaces of vessels containing water. This simple test might seem naïve to those who believed in the textbook wisdom.

According to the current commonly accepted dogma: even if the water of hydration could have different properties from bulk water, the thickness of this zone due to thermal disordering of organized water structures, should not exceed few water layers— several nanometers, at most. Surprisingly Pollack's approach revealed that, in the vicinity of various hydrophilic surfaces, a very thick water layer exists from which solutes are profoundly and extensively excluded.

Under certain conditions, the thickness of this zone reached hundreds of microns. This is equivalent to many hundreds of thousands of water layers. Numerous experiments allowed rejecting trivial and irrelevant explanations for the existence of this "Exclusion Zone" ("EZ water" as it is named by Pollack). He proved that the zone has an aqueous nature, and the properties of water that build it are different in many respects from the properties of bulk water and "usual" aqueous solutions. After years of research, Professor Pollack proposed the concept of a fourth phase or state of water existing beyond our three, well-known states of water: solid, liquid and gaseous.

Numerous experiments revealed multiple peculiar properties of EZ water:

• Solvency of EZ water is much lower than the solvency of the bulk water; it excludes not only colloidal particles, but also high and low molecular solutes that can be easily dissolved in bulk water (Fig. 1).

• EZ water is much more viscous than normal water (about tenfold).

• Water molecules that build EZ-water are much more constrained than in bulk water.

• Water molecules in EZ water are not only constrained, they are evenly oriented: polarizing microscopy shows that EZ water is anisotropic (inhomogeneous) – it has double refraction.

• Unlike bulk water that is transparent for UV-light down to ~200 nm EZ, water exhibits a profound peak of light absorption at 270 nm and emits fluorescence when excited by this wavelength.

All these properties already show that EZ water is physically different from usual bulk water, but two other discoveries made by Pollack, discussed below, are even more significant because they suggest that EZ water may be the source of free energy.

EZ water is charged in respect to bulk water. Its charge depends upon the charge of hydrophilic surface near which EZ water is formed. If the surface is charged negatively, like the surface of hydrogel Nafion carrying fixed sulphate ion groups EZ water is charged negatively with respect to bulk water; if the surface is charged positively, EZ water is also charged positively. However it should be mentioned that most natural hydrophilic surfaces, including the surfaces in biological matter—surfaces of cells, intracellular organelles, nucleic acids, and most proteinatious supramolecular particles—are charged negatively. Depending on the properties of a hydrophilic surface potential difference between EZ water and the neighboring bulk water may reach 150 mV!

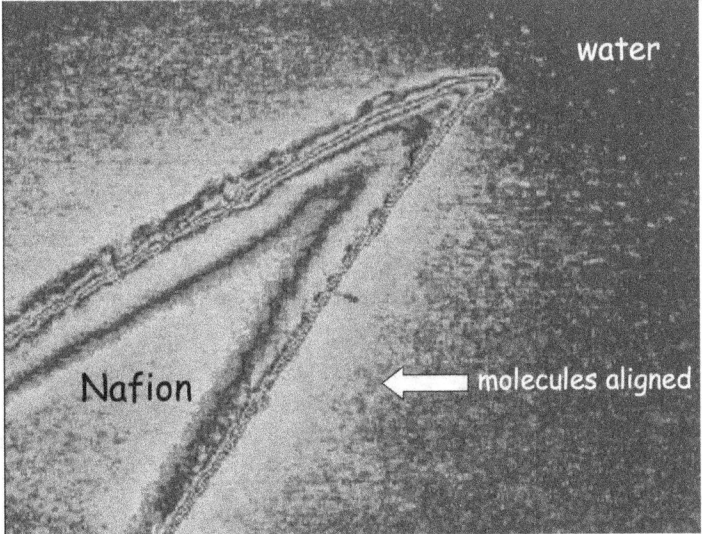

Fig.5. Microspheres are expelled from the Nafion surface at a distance much exceeding 0,2 mm. Clear layer between Nafion surface and microspheres slurry — EZ water.

Electrical potential difference between EZ water can be measured directly. If we insert one microelectrode into EZ water, and another – far away from the surface of Nafion, the voltmeter registers potential difference and ammeter—electrical current, though weak, but not vanishing. (Fig. 6). So pure water acts like an everlasting battery.

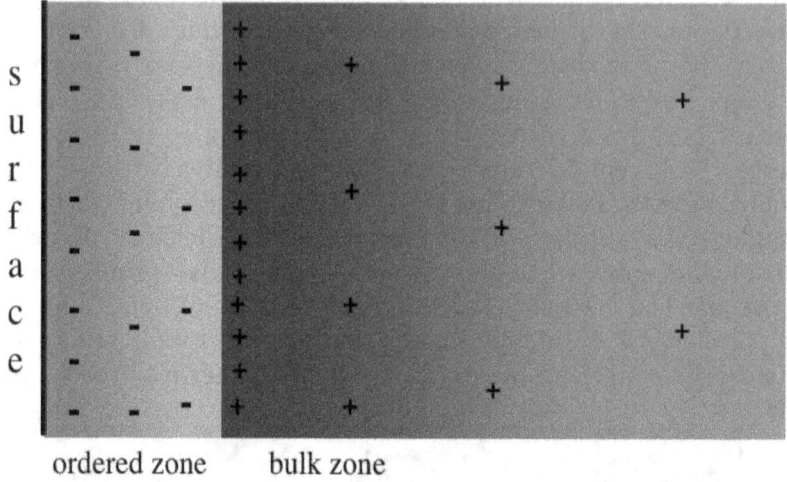

ordered zone        bulk zone

Fig.6. Water battery.

EZ water becomes negatively charged because some water molecules forming it expel protons that are concentrated at the boundary between EZ-water and bulk water. Near the Nafion, the surface pH becomes acidic and, as the EZ water layer grows, the acidic zone shifts away from the surface. As soon as EZ water loses protons, it can no longer be considered as a kind of static entity built up of native water molecules ($H_2O$). Pollack suggests that EZ water is like a kind of a 3-dimensional aquatic polymer fabricated of monomers to which the apparent formula $[(H_3O_2)-]$ n can be assigned.

Pollack's discovery, that charge separation occurs in water with the origination of EZ water, seems to be shocking because it seemingly violates the laws of Nature. In particular, it violates the law of energy conservation: a water battery forms spontaneously and electrical currents between its two poles can flow indefinitely long. What charges the water battery? Pollack has discovered the answer to this question. He demonstrated that EZ reacted to light. When illuminated, the size of EZ increased. (Fig.7). The strongest effect was under the influence of the infrared radiation (Fig.8).

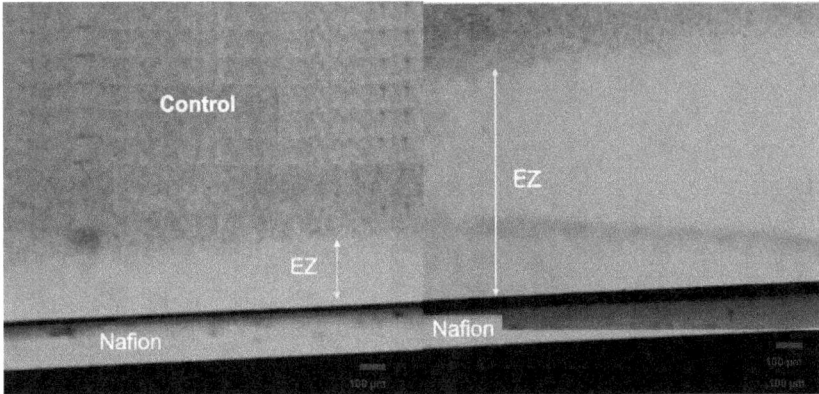

*Fig.7. This is microphotography of the growth of the EZ under the influence of light. Left picture—very dim light, just to make a photo of EZ water; Right picture—the sample, after illumination with visible light.*

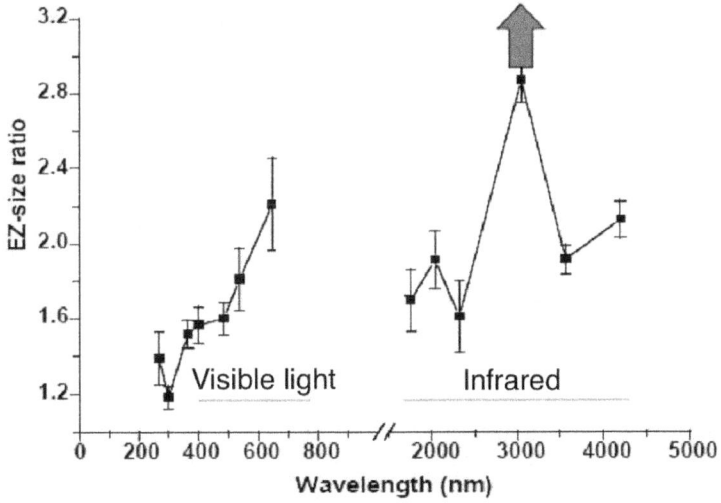

*Fig.8. The dependence of the EZ size upon the wavelength of light.*

This discovery is crucial for understanding the principles of the organization of nature! Before Pollack's discovery, the Sun as the source of visible and IR-radiation was considered to make water more chaotic through heat. It turns out that external radiation may perform quite an opposite effect—it may promote self-organization in water, resulting in the formation of EZ water, leading to a charge separation that turns water into an electric battery! It is especially notable that the most effective inducer of charge separation in water is IR-light.

It is ever-present and everywhere. Its intensity at freezing temperatures produces ice formation and the point where ice becomes liquid is apparently high enough to promote EZ water formation. Thus, we may suppose that, just as liquid water appears, there is a phase separation, in which a charge separation also occurs.

What is equally important to note is that a thick layer of EZ water is also formed at the water/air interface. This suggests that charge separation is not limited to being near solid surfaces. Imagine water moving in eddies and vortices. Besides the phenomena discovered by Walter Schauberger (described above), a huge interface of water and air is formed. It multiplies the activity of water as an electric machine, which increases the capacity of free energy of these dynamically organized entities. The role of potentiation in the preparation of high dilution solutions also becomes clearer. Vigorous shaking leads to charge separation in water. This creates an electromagnetic field in the bulk and contributes to the formation of coherent domains.

Due to charge separation, water may be the source of free energy—energy that may be directly used for the performance of different types of work—electrical, mechanical and chemical. All these types of work may be accomplished in living systems where excellent conditions for EZ water formation are present, and all of them may support vital activity.

But before we turn to the role of water in energetics and bioenergetics, let us briefly summarize the major points of the new vision of water that follow from Pollack's discovery of the Fourth Phase of Water.

The reader may obtain much more detailed information about numerous experiments performed by Pollack and his collaborators and the consequences that followed from his discoveries from his book, *"The Fourth Phase of Water. Beyond Solid, Liquid, and Vapor,"* published by Pollack in 2013. In this work, he provides sound answers to many heretofore-unanswerable questions, for example:

• In the dry sand you can sink ankle deep. But when the sand is moistened with water, it becomes hard—almost like asphalt. How does the water cement the sand grains?

• Waves typically arise and disappear. But how does a huge tsunami come several times around globe virtually without fading?

• Gelatin desserts may be 99.95% water. Why does water not flow away from them? The same applies to the diapers that accumulate fluid: why does it not flow from there?

• You can safely walk on steep dry rocks. But walking takes a lot more skill and attention if the rocks are covered with a layer of water or ice. Why doesn't ice behave like other solids? Why is water so slippery?

In short, the main principles of Pollak's discoveries may be formulated as follows:

### Principle 1: Water Has Four Phases

From childhood, we have learned that water has three phases (aggregate states): solid, liquid, and vapor. Pollack has identified what might be classified as a fourth phase—the exclusion zone. Neither a liquid nor a solid, the EZ is best described as a liquid crystal, with physical properties somewhat analogous to gels—represented, for example, by a raw egg white.

### Principle 2: EZ Water Contains Potential Energy

Water's fourth phase is electrically charged. Commonly, it carries a negative charge, while positively charged hydronium ions concentrate in bulk water adjoining EZ water. Thus charges are separated between bulk water and EZ-water. Since electrons are quazi-free in EZ water, they can be easily donated to the appropriate electron acceptor present in bulk water. Together, charges separated between EZ and bulk waters constitute a battery—a local repository of potential energy.

### Principle 3: Like Charges Can Attract

Pollack and of some other authors have demonstrated in experiments, that like charge particles, (such as beads used for ion-exchange chromatography) suspended in water can attract each other. This is a puzzling observation because it allegedly violates the common knowledge that like charges should repel each other and only opposite charges exhibit attraction.

Experiments performed in water demonstrate the phenomenon of like-likes-like attraction. Pollack suggests that it's not the like charges themselves that attract each other, it's the unlike charges that gather between the like charges that create the attraction. According to him, negatively charged particles create negatively charged EZ water around them; and positively charged hydronium ions that concentrate near this water, attract the negatively charged particles.

However, Emilio Del Giudice and his co-authors recently suggested an alternative explanation of this phenomenon. Rather than of classical electrostatics, their explanation is based upon the principles of quantum electrodynamics. In our opinion, they explain this phenomenon in a more convincing way.

In any case, though many physicists presume that like-like attraction cannot exist, it was predicted by and gained full acceptance through Richard Feynman, a giant of modern physics who, in fact, coined the phrase "like-likes-like."

Even beyond laboratory demonstrations, the attraction may apply more broadly in nature than from biology to geology. The origin of life, for example, almost certainly involves the concentrating of dispersed substances into condensed entities. Without such a condensed mass, no cell could exist. The condensed masses could then become primitive cells. The like-like-likes mechanism provides a simple mechanism for this kind of self-assembly—just add light, wait a bit, and voila!

Another example of like-likes-like occurs in the atmospheric sciences. Consider clouds. Clouds are built of charged aerosol droplets. By conventional thinking, such droplets should repel one another and disperse throughout the sky. But the like-like-likes mechanism provides an explanation for the driving force that unites them together into the entities we recognize as clouds.

### Principle 4: The Sun Imparts Energy to Water (E = $H_2O$)

We don't ordinarily think of water as containing potential energy. A glass of water is considered, more-or-less, in equilibrium with its environment. But the evidence shows otherwise. Water is far from equilibrium—especially when enclosed within the right type of container or when replete with minerals, water will build exclusion zones. Those EZs bear potential energy. Even bubbles, which are surrounded by EZ boundaries, will act as repositories of potential energy. Potential energy is practically everywhere in water. Water constantly absorbs energy from the environment and uses that energy to do work—especially the work of driving flows and movements.

Beyond that obtained from the sun, no energy source is widely and naturally available for doing work as water. The concept may seem a bit less exotic once you consider what transpires in plants. Plants constantly absorb radiant energy from the sun and use that potential energy to do work. Plants, of course, are mainly water. Therefore, it should come as no surprise that the glass of water sitting next to the potted plant transduces incident photonic energy in much the same fashion as does the plant itself. The two are extremely similar.

From a practical point of view, using light to obtain electrical energy from water should also be possible. EZ charge separation closely resembles the initial step of photosynthesis, which entails the splitting of water next to some hydrophilic surface. The similarity is auspicious: if photosynthesis works so effectively, then maybe some other arrangement of water-based artificial photosynthesis has a promising future. Designs built around water might one day replace current photovoltaic designs, which are now seriously depleting the Earth's resources.

At any rate, electromagnetic energy from the sun and other sources of UR radiation creates potential energy in water, which provides an available energy repository. This supply can be harvested for doing work: it can be converted into heat, or it can radiate back out of the system towards the source it came from. The system can handle this gift of energy as needed. The equation $E = H2O$ may suffer a mismatch of units on either side of the equal (=) sign, but it does beautifully capture the essence of the fourth principle: "water stores energy."

The above list of properties appears quite mysterious within the frame of conventional ideas about liquid water. Subsequently, many independent researchers from different countries experimentally demonstrated it. The most concentrated set of numerous experimental and theoretical papers where the new vision of water opening the perspective for the development of the emerging Science of Water is published in the Internet journal "Water" founded by Prof. Pollack in 2007 (www.waterjornal.org). Many dozens of papers were reported at the Annual International

Conferences on *"Physics, Chemistry and Biology of Water"* (www.waterconf.org) that was organized by Professor Pollack in 2004 and has taken place without interruption from then on.

From our point of view, many if not all, anomalies of water discussed above, as well as "miraculous" phenomena, lose their "mystical" ambiance when quite rational explanations—based on the new vision of water that has been emerging to a great degree due to Pollack's discoveries—are applied.

**Water—the transformer of ambient energy into free energy and the working body performing useful work.**

### Incredible mechanical work is performed by water

The knowledge is old and trivial, that flowing and falling water can serve as the source of energy for different devices and, in some cases, may be regarded as the working body itself.

However, here, water is a just a material body that transforms gravitational energy (or energy of external pressure) into work. And it is nearly impossible to imagine that immobile water just sitting in a vessel can also perform mechanical work. However, Pollack not only observed this near-to-miraculous phenomenon, but also gave it quite a rational explanation.

If we placed a thin hydrophilic pipe made from Nafion on the bottom of a cuvette or a glass filled with water, water starts flowing through the pipe and will continue to flow almost ceaselessly. This can be seen using macro spheres suspended in water.

This flow is so surprising because no external force is being applied to it. According to Pollack, the driving force for water flow through the tube originated from charge separation. EZ water covering the walls of the tube expels protons that, already in the form of hydronium ions, $H3O+$ concentrate in its internal core. Unlike viscous EZ water that sticks to the walls of the tube, they are free to move. Being small, like charges, they tend to repel each other and start to move together with some water hydrating these ions outside the tube where they can disperse.

As soon as charge separation in water is provided (by all-pervading external radiant energy), continuous water flows through the tube and has nothing to do with "perpetuum mobile." Rather this system is merely a transformer of one form of energy (electromagnetic) into another (mechanical). This phenomenon may also explain the movement of water through the capillaries of our body, the rise of water within the trunk of a tree, and the movement of water in cracks and pores from the depths of the earth.

Indeed, self-movement of water in narrow tubes, discovered in 1922, may possibly give explanation to the phenomenon of blood circulation in the very early chick embryo before the formation of the heart valves. J. Bremer, who made this observation, described the two streams of spiraling blood with different forward velocities in the single tube stage heart. Nevertheless, the blood had a definite direction of flow within the conduits and moves without an apparent propelling mechanism. In general, this phenomenon supports the very unusual statement that "the heart is not a pump forcing inert blood to move with pressure, but that the blood is propelled with its own biological momentum."

Quite independent of this, reports recently came out that blood movement in capillaries might occur when external pressure is practically absent. Mechanical occlusion of blood vessels in mice ears never result in complete stasis of blood flow in micro vessels: erythrocytes continue to move in capillaries for more than 1 hour after the occlusion.

# Water as the Source of Chemical Work: "Burning of Water"

As noted above, water is the dominant molecule in living matter. The central role water plays in vital activities are generally acknowledged. But usually, when functions of water within a living body are listed, biochemists primarily mention its role as the solvent for the reagents performing biochemical reactions, like transporting nutrients to cells, taking away waste from cells, and as a temperature buffer. Of course, the role of water as chemical reagent in metabolic activity is mentioned as well. In general chemistry, water is considered to be a rather active compound. However, in biochemistry, only water's hydrolytic activity is usually considered. This is the reaction in which water molecule H-O-H splits into fragments (free radicals): OH and H. H is added to one part of the hydrolyzed molecule, and OH – to another.

Hydrolysis of organic molecules is one of the most common biochemical reactions. For example, during the digestion of proteins, hydrolysis of peptide bonds connecting two neighboring amino acid residues takes place and a complex polymeric molecule converts into more simple peptides and amino acids. In order to extract energy from the ATP molecule, it should be hydrolyzed to ADP (accepting H) and phosphoric acid (accepting OH).

Free energy is needed to perform useful work (muscle contraction, for example) and is released in the course of the hydrolysis of the ATP molecule. However, as one can see in Fig. 9, in order to get H and OH for ATP hydrolysis, a covalent bond in the water molecule should initially split. The quantity of energy needed to split a water molecule (about 110 kcal/mol or 4,8 eV per one H-O bond) incomparably exceeds the quantity of energy released as a result of ATP hydrolysis (7-10 kcal/mol).

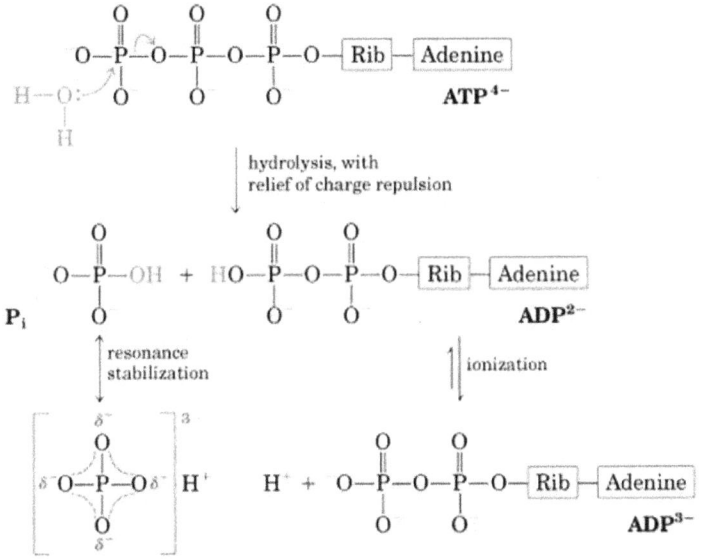

*Fig. 9. Hydrolysis of ATP to ADP + phosphate. Estimated energy output of this reaction ~ 7 kcal/mol.*

Paradoxically, this discrepancy is usually never discussed. Scholars satisfy themselves with the following explanation: that according to thermodynamic balancing, the reactions of hydrolysis are approximately thermo neutral. In the course of recombining the H and OH free radicals with the fragments of the molecule that undergoes hydrolysis, approximately the same quantity of energy is released as is required for splitting a water molecule into free radicals and for a molecule to be hydrolyzed.

Nevertheless, the existence of an overall balance is not the solution to the problem. Energy that might compensate for the energy invested in water splitting only appears after water splitting has occurred and not before.

To our knowledge the only scientist, who suggested the solution to this paradox and who presented experimental evidence supporting his suggestion, was Alexander Gurwitsch (1874 – 1954), the discoverer of the so called "Mitogenetic radiation" (MGR), ultra-weak radiation in the UVrange of the spectrum[24].

Gurwitsch noticed that one of the sources of MGR was the

---

[24] A. Gurwitsch, Physikalisches uber mitogenetische Strahlen, Arch.Entwicklungsmech. 103, 490-498 (1924)

enzyme-catalyzed hydrolysis of different organic compounds occurring in aqueous solutions. Though the intensity of UV-photon flux from reaction systems was very low, the energy of UVphotons could reach 7 eV (equivalent of 170 kcal/mole)–more than enough to split a water molecule and for a molecule to be hydrolyzed.

How could such a quanta of energy emerge in reaction systems, which according to thermodynamic calculation, could generate, at most several kcal/mol? Gurwitsch supposed that the oxygen molecules dissolved in water could be excited, even split into oxygen atoms under the action of incidental visible radiation. Active oxygen atoms, that include free radials, could react with water, thus giving rise to chain reactions in which water could be oxidized and high-density energy, in the form of electronic excitation, could be released.

These suggestions received the rigor of experimental support. Their results found, that if the reaction system became devoid of oxygen or if oxygen activation was obstructed, MGR disappeared[25]. Besides using very sensitive analytical methods, tests proved that hydrogen peroxide–the product of water oxidation–appears in the course of the hydrolytic reaction. This phenomenon seems to be highly unusual in the frame of classical chemical kinetics and thermodynamics and probably explains why the scientific community ignores and continues to ignore Gurwitsch's discovery and explanations.

But now, when we know so much more about the complexity of water, about the presence in any aqueous system, of dynamically organized water phases (Coherent Domains and EZ water, discussed above) that are much more chemically active than bulk water, one may suppose that water belonging to the organized phase may readily participate in oxidation/reduction reactions.

If this is the case: not only does photon emission (including UV photons) accompany allegedly thermally-neutral, chemical reactions become the rational explanation, but many other puzzling properties of water–including such the "miracle" of water burning–may lose the plague of "mysticism." Next, we'll examine, in more detail, the role of water complexity within physical/chemical

---

[25] Voeikov V.L. and Beloussov L.V. From Mitogenetic Rays to Biophotons. In: Biophotonics and Coherent Systems in Biology L.V. Beloussov, V.L. Voeikov, V.S. Martynyuk, (eds.). Springer, 2006, pp.1-16

processes.

The reactions of water, according to the common belief, belong to the realm of exotic inorganic chemistry, participates as an oxidizer, or as a reducer. As general chemistry textbooks teach us, water may play the role of a reducer or oxidizer only if its partner is a very strong oxidizer (like fluorine) or a reducer (like metallic sodium), respectively.

A long time ago, the participation of water, either as a catalyst in the reactions of oxidation, or as a fuel that may be directly oxidized by oxygen, was already described. Many practicing chemists also know this, yet it remains absent from chemical textbooks because no rational explanations to these phenomena in the frame of classical theoretical chemistry can be given.

Let us start from the catalytic role of water in such seemingly simple reactions as oxidation by oxygen of carbon or carbon monoxide—reactions of burning (combustion). The discovery that carbon burning can take place only in the presence of water was made more then two centuries ago. In 1794, British chemist, Elizabeth Fulhame, published an essay on combustion[26], in which she stated that water is the necessary catalyst (or intermediate) of coal combustion. She wrote: "...water is the only source of the oxygen, which oxygenates combustible bodies while hydrogen of water binds to oxygen of air and forms a new quantity of water equal to that decomposed" (cited after[27]). Simple equation:

$$C + O_2 \rightarrow CO_2$$

Describing the burning of coal (carbon) needs to be rewritten according to Fulhame as a sequence of events:

$$C + 2H_2O \rightarrow CO_2 + 4H; \quad 4H + O_2 \rightarrow 2H_2O$$

From such a representation, it only follows that the role of carbon here is to immobilize oxygen from water and to provide for the release of hydrogen (electrons). A real "combustible body" here is water, because it serves as an electron donor for oxygen

---

[26] Fulhame Elizabeth. An essay on combustion, with a view to a new art of dying and painting, wherein the phlogistic and antiphlogistic hypotheses are proved erroneous.– London: Published by the author, 1794.

[27] Laidler KJ, Cornish-Bowden A. (1997) Elizabeth Fulhame and the discovery of catalysis: In: New Beer in an Old Bottle: Eduard Buchner and the Growth of Biochemical Knowledge. Cornish-Bowden A (ed) Universitat de Valencia, Valencia: 123 – 126.

reduction. At the same time, water here plays the role of a catalyst, because the product of oxygen reduction by electrons:

(H = e— + H+) donated by water is again water.

Fulhame's discovery was too soon forgotten. But in 1877, British chemist G. B. Dixon, independent of Fulhame, came to the same conclusion—that water is indispensable for burning. He revealed the necessary role of water in the combustion of carbon monoxide:

$$2CO + O_2 \rightarrow 2CO_2$$

It turns out that a very dry mixture of CO and $O_2$ could not be ignited with a spark unless a drop of water was added to a vessel along with these gases. Even traces of water absorbed on a vessel wall were enough to provide ignition of these gases. Experimental studies continued for more than half a century into the catalytic role of water in combustion[28]. Water was shown to serve as an electron donor for oxygen molecules, while oxygen, thus generated, combined with the combustible body, turning CO into $CO_2$. In spite of the significant efforts of some of the era's most outstanding chemists (M. Traube, D. Mendeleev, and others), the detailed mechanism of the reaction was not established and the phenomenon was again, forgotten.

Although no conventional theoretical foundation for the explanation of catalytic role of water in combustion currently exists, practitioners exploit this property of water. For example, dozens of patents worldwide were issued for devices and methods for burning coal slurrywater suspensions, in which water was present at up to 50% by weight. The inventors and practitioners note that the temperature for the ignition of these suspensions may be hundreds of degrees lower than that needed to set alight dry coal—that combustion goes on much deeper, leaving much less ashes[29].

Water not only plays the role of a catalyst for the combustion of different fuels, it may also literally burn itself. The American inventor, John Kanzius, incidentally discovered that to irradiate a test tube with NaCl aqueous solution (or just seawater) with radio

---

[28] Bon WA (1931). Fifty years of experimental studies of the effect of water vapor on carbon oxide burning. J Chem Soc (London): 338-361.

[29] Manfred RK. Coal-Water Slurry as a Utility Boiler Fuel. Annual Review of Energy. 1986. Vol.11:25-45

waves at 13,56 MHz at room temperature, the water in the solution splits into oxygen and hydrogen. When ignited by a match or a lighter, the solution sustains the flame as long as the water supply lasts (Fig. 10)[30]. Notably, distilled water does not burn under these conditions. This indicates that dissolved salt somehow promotes water splitting under the action of such low-density energy photons as radio frequency photons.

Fig 10. *Flame from NaCl solutions in water (a) 0,3%NaCl; b) 3%NaCl; c) 30%NaCl in Pyrex test tube) after ignition of the solutions irradiated with the beam of 13,56 MHz radiofrequency radiation.*

Such intense water splitting—that could produce enough hydrogen to ignite—is indeed, spectacular. Other scientists also recognized the fact that water can be split by forces of much lower density—by treating it with audible sound—in the course of freezing, thawing, evaporation and condensation, especially when filtration occurs through narrow capillaries[31]. Even stirring water in the presence of the fine powders of NiO, Cu2O, and Fe3O4 has resulted in the release of measurable quantities of hydrogen (and, of course, of oxygen) from water[32].

[30] Roy R., Rao M.L., Kanzius J. (2008) Observations of polarised RF radiation catalysis of dissociation of $H_2O$–NaCl solutions Materials Research Innovations. 12: 3-6

[31] Domrachev A., Roldigin G.A., and Selivanovsky D.A. (1992) Role of sound and liquid water as dynamically unstable polymeric system in mechano-chemically activated processes of oxygen production on Earth. *J. Phys. Chem.* 66: 851-855.

[32] Ikeda S., Takata T., Komoda M., et al. (1999) Mechano-catalysis -- a novel method for overall water splitting, *Phys. Chem. Chem. Phys.* 1: 4485-4491.

One may argue that combustion in a "simple" inorganic system and oxygen utilization in complex living systems (like that responsible for the origination of MGR in hydrolytic reactions) are unlikely to have anything in common. However, an unexpected discovery was made in 2000. All antibodies (immuno-globulins), irrespective of their species and antigenic specificity, and some other proteins (including beta-galactosidase, betalactalbumin, and ovalbumin) could catalyze oxidation of water with singlet (excited) oxygen to form hydrogen peroxide[33].

Since water is the electron donor for oxygen reduction, this process is indeed equivalent to water burning. The immuno-globulin proteins have specific "pockets" (active centers) in which two or more water molecules may be arranged in such a specific configuration, that they obtain reducing properties through their collective interactions (Fig. 11).

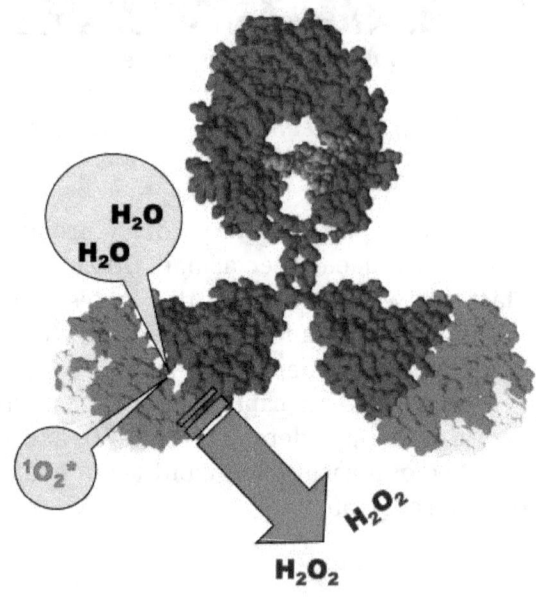

Fig. 11. *A scheme of an Immuno-globulin G (IgG) molecule with the active center, in which water molecules are arranged in such a way that they can donate electrons to an electronically excited singlet*

[33] Wentworth AD, Jones LH, Wentworth P Jr, Janda KD, Lerner RA (2000). Antibodies have the intrinsic capacity to destroy antigens. Proc Nat Acad Sci USA 7: 10930 – 10935

*oxygen (1O2\*) molecule, resulting in H2O2 origination.*

In general textbooks on physical chemistry, a water molecule is considered to be a very poor electron donor. The energy of water ionization is 12.6 eV, which corresponds to an excitation temperature of 145,000°C. However, the evidence mentioned above points to the fact that water can serve as the donor of electrons for oxygen and under much milder conditions than is generally considered. Looking at it from another perspective, this means that water may be more easily oxidized by oxygen (combusted or "burnt") than from the properties of individual water molecules.

One of the major conditions for water to become a reducer, under mild conditions, is that water must contact certain interfaces, including interfaces of some substances dissolved or suspended within it. Besides, water oxidation with oxygen (water burning) also implies oxygen activation. Therefore, water oxidation occurs more efficiently if a specific energy of activation is provided, such as EM radiation in a very wide spectral range–from visible light to radio frequencies.

According to the new conception of water, in most natural aqueous systems, these conditions realize, to a greater or lesser degree, that two different aqueous phases—dynamically organized EZ-like (or Coherent Domains) water and the much less ordered (bulk) water–can coexist.

The major properties of this newly recognized, dynamically structured water described above is in connection with water reducing properties. Let's return to the properties of EZ-water, discovered by Gerald Pollack, to explain its ability to serve as the most peculiar "fuel" for combustion:

• EZ-water has a negative electrical potential with respect to the bulk water adjacent to it (down to -150 mv);
• Protons concentrate at the boundary between EZ-water and bulk water;
• EZ-water has a prominent peak of light adsorption at 270 nm, and it emits fluorescence when excited with this wavelength.
• Thickness of the EZ-water layer increases when illuminated with visible and especially with IR radiation.

All these features strongly suggest that the electrons in this water are much less bound (i.e. they reside at a much higher state of excitation) than electrons in bulk water. Hence, a much lower energy of excitation is needed to make them free. As radiation–especially light in the IR part of the spectrum–increases the thickness of the layer of EZ-water, it also increases its electron-donating capacity. EZ-water becomes a practically inexhaustible source of electrons. Thus EZ-water may be considered as residing in a stable state of non-equilibrium with respect to bulk water, which represents ground-state water.

To convert the potential energy of quasi-free electrons in EZ-water into free energy capable of performing work, an acceptor of these electrons is required. Normally, this acceptor is always available—i.e. oxygen. As already noted, water may split, giving rise to the appearance of free oxygen obtained under very mild conditions. Thus, if EZ water is in contact with bulk water (in which oxygen is dissolved), EZ-water will donate electrons to oxygen. The overall reaction of complete oxygen reduction is represented as:

$$2H_2O \text{ (EZ-water)} + O_2 \rightarrow O_2 + 2H_2O \text{ (Bulk water)} + n*hv \text{ (Energy)}$$

The curious and unique feature of the reaction of water oxidation with oxygen ("burning") is that the reagents on the left side of this equation and the product on its right side appear to be the same (water and oxygen). However, water on the left (in bold) belongs to a stable non-equilibrium (excited) structure—i.e. EZ water. Water on the right side of the equation (in italics) is the ground-state (bulk) water. Besides, water on the left represents a dynamically organized, low entropy phase, whereas water on the right is a much more chaotic, high entropy matter. So, in the course of water burning, water transits from excited to a ground state while at the same time, the overall entropy of the system increases.

In the course of water oxidation with oxygen (or oxygen reduction with electrons donated by water) intermediate products—reactive oxygen species (ROS) such as super-oxide radicals ($HO_2$), hydrogen peroxide ($H_2O_2$) and other chemically active particles, continuously originate and nearly immediately annihilate in reactions of dismutation. A high quanta of energy—equivalent to photons of visible or even UV range of electromagnetic spectra—are released when the acts of ROS annihilation occur. The total quantity of energy released in few acts of annihilation of ROS per every fully reduced $O_2$ molecule, may reach up to 8 eV.

This quanta of energy does not easily dissipate into heat. Instead, it may go on to the electronic excitation of ensembles of molecules present in an aqueous system in which "water burning" proceeds.

Such a scenario fully agrees with Alexander Gurwitsch's ideas of the origin in living cells of "non-equilibrium molecular constellations"—sets of bio-molecules residing in a collective (coherent) state of excitation, including electronic excitation. Such constellations may be regarded as supra-molecular engines responsible for the realization of all the forms of vital activity of a living cell.

According to Gurwitsch, these constellations are kept in the state of excitation due to continuous (or oscillatory) pumping with high-density energy released in the course of oxidative reactions[34]. The only addendum that we suggest to Gurwitsch's concept is the suggestion to consider water burning as the major source of energy used for keeping the molecular constellations within an excited state.

From our point of view, water is not just the most accessible, ubiquitous fuel, but also a unique fuel. A usual fuel will, once burnt, convert into inert "ashes," unable to burn any more. Any "ash" originating from EZ water burning is merely bulk water. Due to omnipresent environmental, electromagnetic fields (represented in particular with IR radiation), this "ash" reverts back to "fuel" and the process of water "burning" may begin anew.

Therefore, the equation of "water burning" presented above should be modified to show that water burning does not generate energy from nothing, it functions as a step up transformer of low grade, low density energy into high grade, high density and organized energy. Such energy can be efficiently used for work performance:

$$2H_2O \text{ (EZ-water)} + O_2 \rightarrow O_2 + 2H_2O \text{ (Bulk water)} + E \textit{ (High density)}$$

$$E \textit{ (Low density)}$$

---

[34] Voeikov V. Mitogenetic radiation, biophotons, and non-linear oxidative processes in aqueous media. In: Integrative Biopysics. Biophotonics. Eds. F.-A. Popp, L. Beloussov. Kluwer Academic Publishers. Dordrecht/Boston/London, 2003, Pp. 331-360.

Due to water's amazing properties (to our knowledge, no other compound behaves like this), it may serve as the inexhaustible source of high grade, high potential free energy that may be used for the performance of different kinds of work—in particular, for work supporting the vital functions of living systems.

Please note, the process of EZ-water "burning" (oxygen reduction by electrons abstracted from the "fuel") as outlined in the equation above, shows an ideal situation that can hardly be realized using very pure water. In fact, for the process of water "burning," to efficiently proceed, certain catalysts are needed. For example, real water "burning," observed by R. Roy et al., under the action of polarized radiation can take place only in NaCl aqueous solutions—not in pure water. Water splitting, under the action of sound and shear stress, results in the elevation of H2O2 to 1-2 orders of magnitude more efficient in MgSO4 solutions than in pure water. Notably, water splitting was negligible in pure degassed water.

Interesting "impurities" that may catalyze the processes related to water splitting and burning perhaps belong to the family of carbonates, including dissolved carbon dioxide, the product of its reaction with water—carbonic acid, that easily converts in bicarbonate, the major representative of carbonates in natural waters, including biological liquids:

$$CO_2 \leftrightarrow H_2CO_3 \leftrightarrow HCO_3^-$$

Carbonates are commonly present in water because of the high solubility of $CO_2$ within it ($CO_2$ is 30 - 35 times more soluble in water than $O_2$) and because of the wide natural occurrence of carbonates in nature. As was recently demonstrated, super-oxide free radicals—the primary product of oxygen reduction (water burning) is continuously being generated in those aqueous systems containing bicarbonate[35]. The intensity of its generation correlates with bicarbonate concentration. It is also remarkable that the rate of oxygen reduction increases upon water illumination with visible light. This agrees with Pollack's observation: that water illumination supports the restoration of EZ water.

The finding that carbonates promote oxygen reduction with

[35] Voeikov V. L., et al. The Stable Nonequilibrium State of Bicarbonate Aqueous Systems. Russian Journal of Physical Chemistry A, 2012, Vol. 86, No. 9, pp. 1407–1415

water is in accord with the long known phenomenon that carbonate species (and especially bicarbonate) play a key role in maintaining aerobic respiration at all levels of living systems organization. It is likely that the ability of carbonate species to catalyze the process of the oxidation of water plays a fundamental role in supplying the bio-energy of vital functions and in keeping living tissue in a stable, excited state, thereby ensuring its extremely high sensitivity to chemical and physical factors of low and ultra-low intensities.

It's also important to note that processes of EZ-water "burning" (oxygen reduction by electrons abstracted from the "fuel") or the "burning" of other substances in water, like any other burning, should proceed as a branching (avalanche-like) chain reaction, which must obey particular laws pertaining to such processes[36].

Combustion can only begin if oxygen concentration exceeds a certain threshold. It is initiated when some triggering stimulus whose energy may be incommensurably smaller than energy released in the course of the reaction's development. Any energy released should further promote the excitation of EZ water and oxygen, resulting in a reinforcement or invigoration of the burning process.

When the availability of either oxygen or electrons falls below threshold levels, burning is dampened. However, so long as hydrophilic surface-organizing EZ water exists, water molecules (from the bulk) may be recruited to the EZ water. In addition, part of the energy released in the course of burning inevitably degrades as heat—the IR-part of spectrum—which may induce a further increase of EZ water and its potential, i.e. its stock and level of excitation of quasi-free electrons.

During this period, oxygen—a product of the reaction of water burning—again accumulates, and a new wave of water "burning" may arise. Thus the process could become oscillatory. In turn, energy will be released in an oscillatory manner and may serve as a pacemaker for coupled reactions.

Fig. 4 illustrates the reaction of water burning that was

---

[36] Voeikov VL, Naletov VI (1998). Weak Photon Emission of Non-Linear Chemical Reactions of Amino Acids and Sugars in Aqueous Solutions. Evidence for Self-Organizing Chain Processes with Delayed Branching. In: «Biophotons». Jiin-Ju Chang, Joachim Fisch, Fritz-Albert Popp. Kluwer Academic Publishers. Dortrecht, The Netherlands, pp. 93-108

triggered by a 5-min irradiation of bi-distilled water with infrared laser ($\lambda$ = 1264 nm, power 5 mW). Photons of this wavelength induce the transition of dissolved oxygen to the electronically excited ("singlet") state that may "ignite" same kind of organized water. After a prolonged lag-period (a feature characteristic for the avalanche chain reactions), water luminescence in the blue-green region begins to flare up and the process takes the shape of auto-oscillations. Such luminescence of water lasts for many hours. The periods of pulsation were of about 300 and 1150 seconds. It is notable that hydrogen peroxide—the intermediate product of water oxidation—accumulated in water from which photon emission was registered.

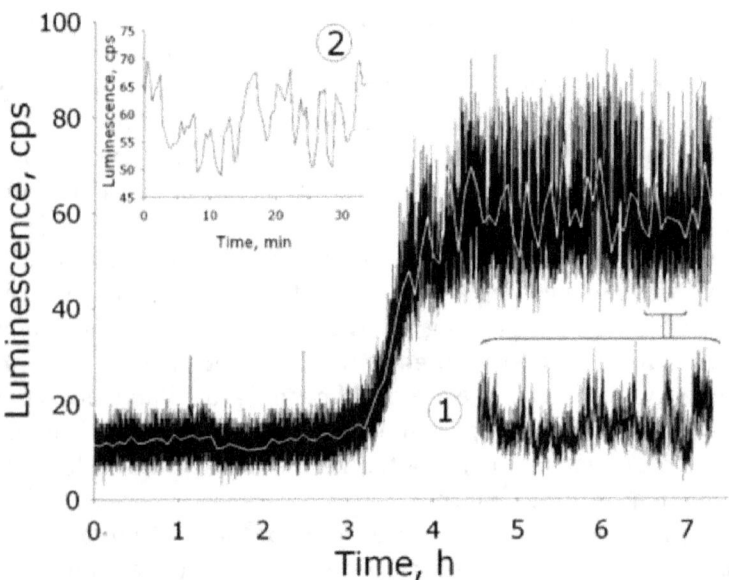

Fig. 12. *The effect of laser radiation on water luminescence. Water was illuminated with infrared laser for 5 min (a vertical arrow). To the left of the arrow is a record of the background water luminescence before the onset of laser irradiation. The white line on the basic plot presents the macrostructure of a signal obtained using the curve smoothing. Inset 1 shows the microstructure of changes in water luminescence. In inset 2, the integral intensity of water luminescence corresponding to that in inset 1 is shown.*

What kind of external work may be performed by EZ-water for

self-sustaining and increasing the total stock of the system's structural energy? If $CO_2$ and $N_2$ are present in such an aqueous system, and they are always present in natural waters on Earth, then energy of electronic excitation released in the course of oxygen reduction can be used in their activation.

Due to the reductive potential of EZ water, carbonyls and amines may be produced. This allows for the initiation of chains of chemical reactions known as amino-carbonyl reactions in which complex organic molecules arise and polymerize. Hydrophilic polymers and their assemblies arising in water, present new hydrophilic surfaces that will transform more and more bulk water into EZ water. Consequently, in such an evolving aqueous system, a process of self-organization takes place. The self-organization process increases its stability to the action of external forceful factors while, on the other hand, promoting its sensitivity to the action of weak and resonant factors. Ultimately, the system may perform more and more external work promoting its sustainable development—the major feature of the progressive evolution.

As already noted, the process we've discussed proceeds according to the principles of branching chain reactions. Regarding such reactions, it is interesting to recollect the proposition of Sir Cyril Hinshelwood who, together with Nikolai Semenov, was awarded the Nobel prize for the discovery of branching chain reactions: "... It is very possible and even rather probable, that from the very instant of life's emergence on the Earth, a giant branching reaction has taken place."

Branching chain reactions in chemistry are the reactions of combustion, and are usually regarded as destructive. However Hinshelwood prophetically predicted that this principle might underlie the process of constructive, progressive evolution of life on the Earth. Indeed, as we tried to show in this chapter, in natural aqueous systems in which water burning continuously occurs, progressive evolution should keep up with necessity. Due to the unique properties of water, and only of water, being able to recover back into fuel (EZ water) from the ashes of its predecessor—like the mythological bird, the Phoenix, this constructive burning may continue for as long as we can foresee.

# The Science of High Dilutions

*Facts, unexplained by existing theories, are the most dear to science, their development would guide the development of science in the near future.*

A. M. Butlerov,
Russian chemist (1828-1886)

The hypothesis that the action of biologically active substances in ultra-low doses is mediated by the changing of water, in which the studied processes are implemented, even backed up with theoretical ideas about the dynamic structures in water in accordance with the QED principles, required experimental confirmation. But conclusive evidence for this did not exist almost until the beginning of this Millennium.

But with the onset of a new era — the era of "Aquarius," experimental evidences of unexpected changes in the physico-chemical properties of water with ultra-low concentrations of biologically active substances, or subjected to the action of physical fields of ultra-low intensity, began appearing as from cornucopia. Scientists of different specialties, working in different countries, almost simultaneously obtained these results. Taken collectively, their results open possibilities. It may yet be possible to explain, on a strictly scientific basis, all the phenomena, heretofore rejected because of their "unscientific background" from the narrow perspective of the classical concepts of physics and chemistry.

A. I. Konovalov resolved many of the issues raised in the works of E. B. Burlakova. Alexander Konovalov is an outstanding Russian scientist and organizer of science. For many years, he was the Rector of Kazan University. For ten years, he headed the Institute of Organic and Physical Chemistry of the Kazan Scientific Center of the Russian Academy of Sciences. All this time, he has conducted extensive scientific work.

Konovalov is the author of nearly 1,000 scientific publications. He has more than 50 patents, some of which are implemented in the industry. He also supervises the training of graduate and doctoral students and chairs numerous scientific and editorial boards. In 2014, colleagues and friends congratulated him on his

80th birthday. It's not often you can meet a person with such energy and high intellect combined with deep intelligence and kindness. We echo wishes for his good health with hopes that many years of creative life still stand before him as well as Aleksander Ivanovich.

In 2006, A. I. Konovalov initiated work on the study of the physico-chemical properties of highly diluted solutions. Over the years, they have studied the solutions of dozens of different substances. The most advanced complex of physico-chemical methods were applied: dynamic light scattering, micro-electrophoresis, conductivity, tensiometry, pH-metry, dielctrometric, polarimetry, atomic force microscopy, UV-spectroscopy, and EPR spectroscopy. All these methods account for the complexity of the study. Comparing the results of a combination of methods provides greater opportunities to derive better conclusions than only having the information of only one method.

Solutions were prepared in the classical way—by consecutive dilution of the initial substance in bi-distilled water, accompanied by obligatory shaking after each dilution, then letting it settle in for several hours. At each stage of dilution, physico-chemical properties of the solution were determined. Very important — its biological activity, such as the degree of activation or inhibition of protein kinase C by the action of solutions of antioxidants were also determined.

The series of investigated compounds include: antioxidants, plant growth regulators, neurotransmitters, vitamins, tranquilizers, hormones, various drug substances, as well as substances of unknown biological properties. From the chemical perspective, this list presents compounds with different structures—from simple molecules (e.g., glycine, the simplest amino acid) to complex macro-cyclic-types, such as compounds of porphyrins or calixarenes. Widely used in modern chemistry, Calixarenes are macro-cyclic compounds based on phenols.

The main result obtained by the Kazan scientists in their studies of the physico-chemical properties of aqueous systems with high dilutions, was the fact that, in most cases, the conductivity, pH values, and the values of the surface tension of the preparations varied. In a series of sequential dilutions in a non-monotonic way, it increased, and then decreased.

Already this was surprising. One would expect that decreasing the concentration of the substance due to serial dilution with bi-distilled water, all these parameters would gradually reach the values characteristic of bi-distilled water. The most striking results were obtained using the method of dynamic laser light scattering (fig.13). In this method, a laser beam is directed at a thin layer of water and the resulting pattern of reflected light is analyzed. If the water is uniform, the scattering pattern would be a uniform field of illumination. However, in the presence of inhomogeneities, (i.e., particles which scatter light), glowing spots appear in the picture and allows us to estimate the size of the inhomogeneities.

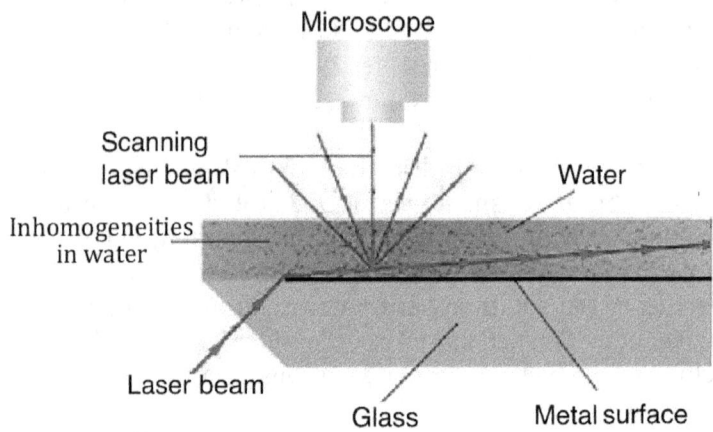

*Fig.13. The scheme of laser dynamic light scattering.*

Aqueous solutions of various compounds, in serial dilution with high-purity water, form light-scattering nano-scale molecular ensembles, despite the fact that water used for dilution is absolutely homogeneous. Konovalov named these molecular ensembles "nano-associates." Their size changed with dilutions in nonlinear and nonmonotonic ways (as, indeed, did all other physico-chemical parameters of the serial dilutions).

The nano-associates in some cases measured up to 400 nanometers—a tenth of a micron—in size. If the nano-associates appeared in concentrations below $10^{-9} - 10^{-10}$ M, they did not disappear after subsequent dilutions (ranged only their average size), at least up to the dilutions corresponding to the calculated concentrations of $10^{-20}$ M and below.

Thus, their size is thousands of times higher than predicted by quasi-geometric models. Long-lived aggregates of water molecules united by hydrogen bonds, can range, more or less, from several dozens to several hundreds water molecules with lifetimes no more than a few microseconds. The nano-associats consist of millions and tens of millions of water molecules. Their lifetimes are days and weeks. Their existence cannot be explained on the basis of standard models of local interaction of the molecules.

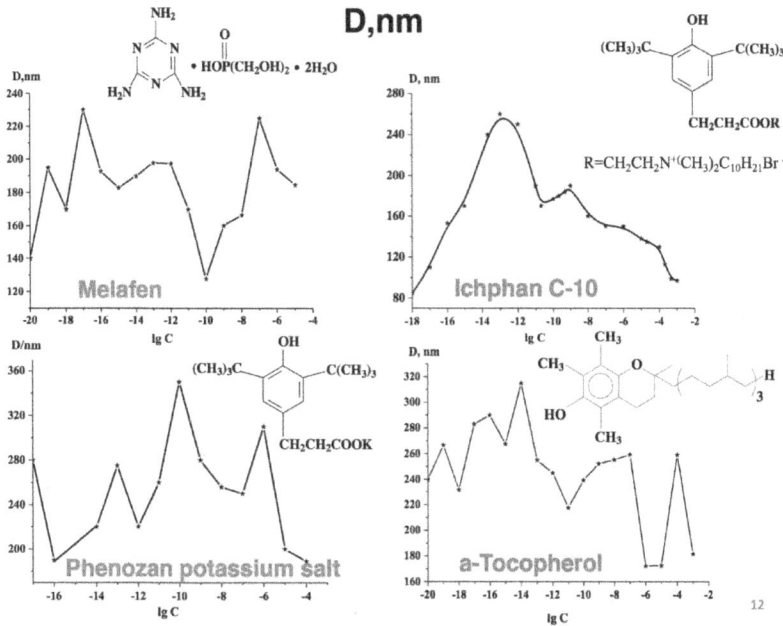

*Fig.14.   The concentration dependence of the size (D) of nano-associates formed in solutions of different substances.*

In his numerous works, A. I. Konovalov presented his experimental curves of concentration dependencies of physical-chemical properties and biological effects of different solutions. They all possess an oscillatory character. First, the activity increases dramatically at certain concentrations. Then, with subsequent dilutions, activity disappears and sometimes transforms its sign to the opposite. Then, increases again. And so it continues throughout the full range of investigated concentrations, of up to 10⁻²⁰ M. The peaks of activity of different substances are not the same—obviously determined by the properties of the original substance.

Moreover, not all substances demonstrate this property. On this basis, Konovalov divided them into substances with "classical" and substances with "non-classical" behavior. Among the investigated compounds, about 25% demonstrated "classical" behavior, while 75% behaved "non-classically." The "classic" behavior is characterized by a quite rapid achievement of the properties of a solvent after a series of dilutions with no further changes of the solution properties. In this case, both the surface tension and electrical conductivity at concentrations of $10^{-6}$ – $10^{-7}$ M have reached values of water, which did not change with further dilutions.

In control experiments, when water was diluted with water without any substance, no effects were observed. According to Konovalov: "No substance—no effects." These results are of crucial importance and not only for medicine and biology. It is obvious that the results obtained open up vast prospects for practical applications. A broad perspective of highly diluted solutions applications in agriculture may drastically reduce the need for nutrients and fertilizers, which would reduce the burden on the environment.

At the same time, it is clear that the time for intuitive development ends. The brilliant insights of Hahnemann must be translated into modern scientific language. Using Konovalov's and the approaches of other researchers, we can create a fully scientific basis of homeopathy. It's such a pity Jacque Benveniste could not live to see this. The next discovery of A. I. Konovalov demonstrates the dependence of homeopathic effects on physical, even cosmo-physical factors.

His experiment was conducted as follows: a prepared solution was divided into two parts (at each concentration); one sample was left, as usual, on the table (aptly designated the "lab table" for this series); the other sample was placed for 24 hours into the permalloy container. Inside the permalloy container, the sample was fully shielded from external electromagnetic fields. The results were amazing. In the "permalloy container" series, no effects were observed at concentrations below $10^{-6}$ – $10^{-7}$ M. It was clean water.

Thus, a weak geomagnetic field plays a crucial role in the effects of high dilutions. This opens up a wide area of research. One can study the intensity and frequency of the influencing electromagnetic field and its dependence on the duration of the field influence to the solution. Obviously, the main component is the electromagnetic field of the Earth, because similar effects were observed for over two hundred years, when people were just beginning to get acquainted with electricity and no man-made sources of electromagnetic fields yet existed.

Thus, the reason for the impact of weak environmental factors on the biological world— change of seasons and moon phases, not to mention solar activity—becomes intuitively clear. We are sure that within our body, many processes take place with participation of high dilutions. Therefore, the resulting data is not only relevant to homeopathy, but to the principles of the organization of biological life.

Let's remember, that the beginning of modern electromagnetic field (EMF) theories for biology began in 1970 with the groundbreaking work of Russian scientist, Alexander Presman. He argued that environmental electromagnetic fields have played a large, if not central, role in the evolution of life and are further involved in the regulation of the vital activity of organisms. Living beings behave like specialized and highly sensitive antenna, systems of the diverse parameters of weak fields comparable to the strength of the ambient natural ones.

According to Presman, electromagnetic fields play an important role in the communication and coordination of physiological systems within living organisms. They also mediate the interconnection between organisms and the environment. However, only recently have bio-electromagnetic field theories reached a certain maturity, due to the general progress in electromagnetic theory, bio-electromagnetic theory, quanum theory, and in particular, non-equilibrium thermodynamics. Mainly, this was achieved in the work of Herbert Froehlich and what could be called the Froehlich School of Biophysics, as well as in the quantum field theoretical approach of Umezawa, Preparata, Del Giudice and Vitiello.

Thus, numerous experiments demonstrated that solutions with "non-classical" behavior form nano-associates, and their formation is the reason for this "non-classical" behavior of the solutions. In solutions with "classical" behavior, nano-associates are not formed. Therefore, when there are no nano-associates, there are no "classical" behaviors of the solution; when there are nano-associates, "non-classical" behavior results. In clean water, nano-associates are not formed. Thus, no initial substance in the solution equals no nano-associates.

All discoveries by Konovalov featured solutions directly relating to their biological effects. E. L. Maltseva, V. V. Kasparov and N. P. Palmina (Institute of Biochemical Physics after N. M. Emanuel of Russian Academy of Sciences) investigated the change in microviscosity of lipid components of membranes under the influence of potassium phenozan solutions in the "lab table" and "perm-alloy container" conditions. It turned out that in the "laboratory table," case effects appeared at three concentrations: $10^{-6}$, $10^{-12}$, and $10^{-15}$ M. In the "perm-alloy container" samples, a preserved effect was observed at $10^{-6}$ M, while the other two disappeared.

The research identified an interesting connection of bio-effects linked with the size of nano-associates. The bio-effect at $10^{-6}$ M corresponds to the maximum of nano-associates' sizes. The bio-effect and the corresponding nano-objects in conditions of "perm-alloy container" are retained, but with further dilutions, disappeared—the solution turned into ordinary water. Bio-effects at $10^{-12}$ and $10^{-15}$ M, formed under the influence of electromagnetic fields, corresponded to the nano-associates of minimum size. Hence, the first conclusion was as follows: the nature of the effects at $10^{-6}$ M and at $10^{-12}$ M, $10^{-15}$ M was different.

Thus, the experimental evidence indicates that the main role in biological and physicochemical experiments do not belong to the individual molecules of biologically active substances, but to their spatially organized nano-associates, which are formed with the participation of water structures. Nano-associates—due to their highly diluted aqueous solutions—are characterized by a certain set of physico-chemical and biological properties significantly different from the properties of pure water.

The question arises whether it is possible, on the basis of the obtained results, to make predictions about the possibility of the existence of bio-effects in highly diluted aqueous solutions under normal conditions? Based on the known physic-chemical properties of such solutions, some ideas may already be formulated. If the solutions behave "classically," bio-effects will not happen. If they possess "non-classical" behaviors, we can identify areas of concentrations where effects can be expected. For more than ten cases, Konovalov and his team have made such predictions.

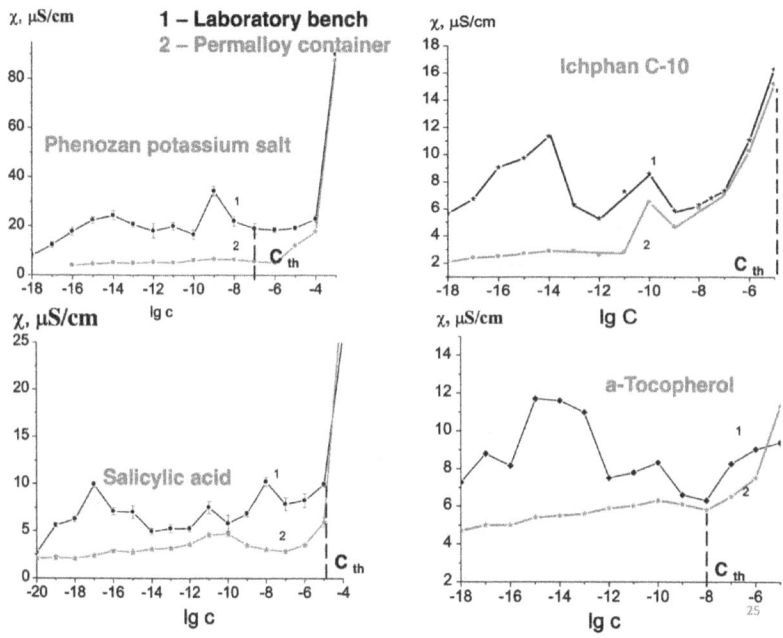

*Fig. This 15. The concentration dependence of the conductivity of solutions of different substances from the "lab table" and "perm-alloy container" series.*

But what do we know about them? What is their composition? Be aware that the sensitivity used in experiments with laser scattering in water to determine the size of nano-objects in solutions, requires at least one thousand nano-objects in one milliliter of solution. Estimates show that nano-associates—already at concentrations of solutes $10^{-10}$ - $10^{-11}$ M—are almost entirely composed of water molecules, and at higher dilutions they consist only of water molecules. We must keep in mind that the nano-crystal of 100 nm size corresponds to about 7 million molecules of water, 200 nm ~ 50 million, 300 nm ~ 200 mn, 400 nm ~ 500 million water molecules.

Naturally, the question arises: how can we explain the formation of nano-associates? The classical model of the electrostatic (Coulomb) interactions between the substance and the solvent does not explain the emergence of large sustainable structures. This explanation was revealed in the concept of coherent domains developed by Emilio Del Giudice, based on quantum electrodynamics.

Konovalov invited Israelian physicist Tamar Yinnon to conduct the calculations[37], which showed that, in the presence of an electromagnetic field (due to the interaction between the plasma oscillations of the dissolved molecules and water associates), several types of coherent domains (CD) with a certain quantum parameters may be created:

CDelec =      ~102 nm
CDplasma = ~ 103 nm
CDrot =      ~ 104 –105 nm
CDrot =      ~ 104 –105 nm

This is another crucial result of A. I. Konovalov and his team. The experimental data showed full compliance with predictions of the quantum electrodynamics theory. It's a pity that Emilio Del Giudice did not see these results...

Research in the field of high dilutions is actively developing in the world. The Internet provides great help in the dissemination of information. In particular, among the myriad of online journals, we can examine the *International Journal of High Dilution Research* http://hydrometeorology.ws/str42.html. In this edition, the scientific articles published are mostly devoted to homeopathy, although there are also interdisciplinary research papers. Please note, Professor V. L. Voeikov is a member of the editorial Board of this Journal.

We are honored to receive and publish a paper Professor A.I. Konovalov co-wrote specifically for this book. From his many years of intensive research, we proudly present Professor A.I. Konovalov and I.S. Ryzhkina's brief summary of his main results, *"The Science of Nano-associates."*

---

[37] Yinnon TA, Liu ZQ, Domains Formation Mediated by Electromagnetic Fields in Very Dilute Aqueous Solutions: 1, 2, 3// Water, 2015, 33-47, 48-69, 70-95.

# THE SCIENCE OF NANO-ASSOCIATES

A. I. Konovalov and I. S. Ryzhkina

*Every new discovery poses more problems than it solves.*

de Broglie

Since the year 2006, a research team from the A.E. Arbuzov Institute of Organic and Physical Chemistry at the Kazan Scientific Center of Russian Academy of Sciences, headed by academician A.I. Konovalov, has been studying what seems to be an unusual phenomenon: the emergence of biological effects of bio-systems under the action of aqueous solutions with "ultra-small doses" or "ultra-low concentrations" of solute. In Russia, this phenomenon has been discovered and studied using instrumental methods (we emphasize this specifically) by the E.B. Burlakova's scientific school in the N.N. Semenov Institute of Chemical Physics RAS, and later in the N.M. Emanuel Institute of Biochemical Physics RAS[38].

An example of such phenomenon is data that demonstrates a change in protein kinase C enzyme's activity under the influence of a successive series of potassium phenozane solutions, "produced by successive decimal dilution of the initial solution, prepared from a sample weight of the original substance with a compulsory shaking after each dilution." (The phrase within quotation marks is hereinafter referred to as an "accepted procedure" or just a "procedure" of the solution's preparation.

Sometimes works of this kind utilize centesimal dilution instead

---

[38] Burlakova E.B. Biological effects of ultra low doses. Vesti RAN, 1994. 64. N 5. C. 425. (in Russian)

of the decimal one. However, in this work, only the data regarding the decimal dilution will be considered). In accordance with the accepted procedure, only the first spot has the indubitable concentration value (in M). The concentrations of the rest are theoretically calculated in concordance with the ratio of the initial solutions' successive dilution. It is important to note that it cannot be done otherwise.

It is impossible to experimentally obtain the indubitable values of concentrations in these quantities, for example, of $lgC = -15$, $-18$, $-20$. One can only calculate them. However, in no way does it affect the principal conclusions. For the highly diluted aqueous solutions, the calculated concentration values are more likely to be overestimated than underestimated, due to the possible solute's adsorption with the successive dilutions.

From the obtained data, we noted two bursts of activity from the protein kinase C solutions: at $lgC = -6$ and at $lgC = -18$. The unusual nature of the phenomenon lies in the fact that although the effect observed at $lgC = -6$ does not raise questions within scientific society (it is located in the so-called therapeutic concentration field), according to the nature of current concept of solutions, there should be absolutely no effect at the $lgC = -18$ (in this case, the concentration is too low, i.e., there is not enough molecules within the solute to lead to a bio-effect). That is exactly why the authors of the work called the observed phenomenon to be the effect of "ultra-small doses" or "ultra-low concentrations," which suppose the systems under study are solutions. But under certain conditions (ultra-low concentrations), there emerges a phenomenon of an unknown nature (please, not this) when they act upon bio-systems.

We confess to being faced with a scientific contradiction: if the studied aqueous systems are indeed solutions, then, for the forgoing reasons, no bio-effects should be observed. Nevertheless, they exist! This fact has been repeatedly, conclusively and instrumentally proven. The claim of the emergence of a phenomenon of an "unknown nature" does not make the situation any clearer. Instead, it raises doubts regarding the reliability of the experiment.

But, perhaps, everything is much simpler: the studied aqueous systems [water + substrate (note: we are beginning to avoid using the term "dissolved substance")] are not solutions in a conventional sense. Additional research is required to establish the truth, in which the classical procedures should be combined with novel methods. And such research has been conducted.

Taking place is a complex, large-scale systematic physical-chemical study of aqueous systems ("solutions" by name, but not by the nature) of different substances with different chemical natures (more than 100 compounds) in a broad interval of concentrations (ranging from $10^{-2}$ to 10-20 M). The methods used include dynamic and electro-phoretic light scattering (DLS), nano-particle trajectory analysis (NTA), conductometry, tensiometry, viscosimetry, pH-metry, atomic force microscopy (AFM), and transmission electron microscopy (TEM)[39].

As a result, a previously unknown fundamental phenomenon has been discovered[40] (experimentally, not hypothetically, 2009) [2], which consists of the formation of nano-sized substrate-induced molecular ensembles with participation of water molecules, called "nano-associates," in the above-mentioned aqueous systems ("solutions").

Such nano-associates are described by the following parameters: their hydrodynamic diameters (D) range from 100 to 400 nm, and their $\zeta$–potentials–from -2 to -20 mV. These parameters change non-monotonically while the systems under examination undergo a process of decimal dilution. The dynamic pattern of the nano-associates' parameter changes is individual for each substrate (is programmed by the substrate) and corresponds to the emergence of the bio-effects in concrete systems, i.e. the nano-associates are the carriers of the substrate's molecular information.

The nano-associates' parameters—their reasonably large size (of course, on a nano-scale) and the fact of $\zeta$-potentials presence—point out that nano-associates are a phase. Their parameters are

---

[39] A. I. Konovalov and I. S. Ryzhkina, Highly Diluted Aqueous Solutions: Formation of Nano-Sized Molecular Assemblies (Nanoassociates)// Geochemistry International, 2014, Vol. 52, No. 13, pp. 1207–1226.
[40] I.S. Ryzhkina. L. I. Murtazina, Yu. V. Kiseleva, and Academician A. I. Konovalov, Supramolecular Systems Based on Amphiphilic Derivatives of Biologically Active Phenols: Self-Assembly and Reac- tivity over a Broad Concentration Range// Doklady Physical Chemistry, 2009, V 428, 2, pp. 201–205

different from the medium, i.e. the studied aqueous systems are nano-heterogeneous, and not homogeneous (what they would have been in the case of solutions). Therefore, they cannot be solutions in the strict sense of this term. They are nano-disperse systems. As a reminder, disperse systems are heterogeneous systems consisting of two or more phases with a developed interface surface between them. One of the phases usually makes for a continuous disperse medium, in which the disperse phase is dispensed. In our case, the role of the disperse phase is played by the nano-associates, which correspond to the respective dilution rate of the system, while the role of the dispersion medium is played by the solutions, also corresponding to the dilution rate, and, finally, water.

This very conclusion is of great importance for the contemplated problem. If such aqueous systems were, indeed, solutions everything would have been plain to see: such a phenomenon cannot occur, and the discussion must be ended.

However, if they are actually nano-disperse systems, we should not arrogantly abandon the facts already achieved from the experiment. Instead, we should re-investigate them and not trust blindly that this time, new results will rain down as if out of the horn of plenty. Nonetheless, thorough results have already been obtained, although a considerable number of questions also arise. That is hardly a surprise, since, as the great de Broglie used to say: "Every new discovery poses more problems than it solves."

The following conditions have to be met in order for nano-associates to form:

1) A special structure of the substrate (although we do not know exactly which yet) is required. Nearly 25% of the examined compounds do not form nano-associates.

2) (This is an extremely significant observation, and can be called "a discovery within a discovery"). Starting from a defined value of the dilution rate (so-called "threshold concentration" ranging from $10^{-5}$ to $10^{-8}$ M), the presence of an external low-frequency electromagnetic field (EMF) is also required. Nano-associates do not form if the examined samples are shielded from EMFs, e.g. if they are being held for 20 making it possible to draw a well-grounded conclusion, outlined in the following paragraph.

3) An above-mentioned procedure of sample preparation.

The formation of nano-associates has been shown to be

responsible for the abnormal physicochemical properties (different from those of water) that aqueous systems display at high dilution rates. In this case, the fact that nano-associates are also responsible for the emergence of their biological properties (the ability to display bio-effects) is of primary importance.

When no nano-associates are formed according to the above-mentioned conditions (e.g. when keeping the prepared aqueous systems in the perm-alloy container), then neither the abnormal physicochemical, nor the biological properties are present in such systems. Thus, if there are no electromagnetic fields present, there are no nano-associates.[41] Hence, such systems display neither the abnormal physicochemical properties of water (everything matches the initially used bi-distillate) nor any bio-effects, although the admissible concentrations of the studied substance exist (!).

Therefore, everything matches the behavior of solutions. It means that "ultra-small doses" or "ultra-low concentrations" have absolutely no effect. The phenomenon is observed strictly subject to the presence of the nano-associates.

It is perfectly clear that the substrate's molecules allocate between the nano-associates (the disperse phase) and the dispersion medium (the solution), although the distribution ratio remains unknown yet. With dilution, the substrate's content decreases both in the nano-associates and the medium, reaching extremely low quantity and even a full absence of its molecules (for example, by lg C = -25).

However, the experimental facts show that presence of the nano-associates is registered even under such conditions. This means a disperse system of a "water in water" kind emerges. The role of the disperse phase is played by the nano-associates that

---

[41] I. S. Ryzhkina, L. I. Murtazina, and Academician A. I. Konovalov, Action of the External Electromagnetic Field Is the Condition of Nano-associate Formation in Highly Diluted Aqueous Solutions// Doklady Physical Chemistry, 2011, V. 440, 2, pp. 201–204
Konovalov A., Ryzhkina I., Maltzeva E., Murtazina L., Kiseleva Yu., Kasparov V., Palmina N., Nano- associate formation in highly diluted water solutions of potassium phenosan with and without permalloy shielding// Electromagn. Biol. Med., 2015. 34(2). pp. 141–146

consist of the water molecules, which are in turn organized in accordance with the program specified by the substrate and carry its molecular information. A simple calculation shows that the size of such nano-associates denotes a contribution of hundreds of millions (!) water molecules to their formation.

Thus, by formation and subsequent transformation of the nano-associates (while diluted) the aqueous systems "read, store and reproduce biologically important molecular information of the substrate" (these words are taken from a letter written by a prominent Italian scientist Foletti). The nano-associates are indeed the carriers of the substrate's molecular information, even under such conditions when the substrate's molecules are themselves absent (high dilution). It is worth noting that the formation of the nano-associates has been demonstrated through the example of several homeopathic systems.

It seems natural that the question regarding the nature of the nano-associates has appeared at the very beginning of the research. Until this moment, nothing has been said about this topic, and for a good reason. The objects are very complex: they consist of hundreds of millions (!) of molecules. They are affected by external electromagnetic fields (EMF). It is perfectly clear that they cannot be described using simple models or simplified methods. An insightful approach from theoretical physicists is required. Specialists who work with problems of water structures (both pure and with the involvement of added substrates) have undertaken such attempts in the sphere of quantum electrodynamics (QED). According to the obtained data, if there is external EMF presented, a formation of nano-sized coherent domens (CD) of different types, depending on the dilution rate, takes place in such systems. These results are promising. However, to be fair, there also are other views regarding this phenomenon.

We think that we have laid the foundation for the science of nano-associates, which, we believe, is the science of the future. Indeed, as of yet, we don't know how widely the nano-associates are spread, and the role they play in the functioning of the living systems also remains unknown. However, the fact that the phenomenon described in this essay should continue to develop is beyond any argument. And following the great Russian chemist, A.M. Butlerov, one can exclaim: "How much work is up ahead, what room for an inquisitive mind!"

# Water and Biological Structures

In previous chapters, we have considered the views on water from the perspective of physicists, chemists and theorists. All their obtained results collectively provide a foundation to introduce the concept of "information capacity" and the "memory" of water.

Under water memory, we understand the process of information transfer to the water system, the formation of a specifically organized ensemble of coherent domains, and preservation of this structure for an extended time.

The change of the physico-chemical properties of water and its biological activity under the influence of low-energy and low-intensity factors continue for a period of time much longer than the lifetime of the hydrogen bonds within water molecules.

From this point of view, we can consider a variety of "healing" waters—Epiphany, "charged," structured, and the like. Only now can this question be investigated without reprisal. Using a variety of modern techniques, the truth can now be differentiated from wishful imagination to outright deception.

But there is another aspect of the water problem that attracts more and more attention by both researchers and the general public. We are talking about the impact on living organisms—particularly human—of small and ultra-low doses of biologically active substances (BAS), low doses of ionizing radiation, light, electromagnetic and magnetic fields of different frequencies, and other physical fields. From our point of view, the reaction of water is the basis of all such effects because our body contains 70-80% of liquids, and all the basic physiological processes occur in the liquid phase.

Under low and ultra-low doses of physical factors and the level of impact on the biological object is negligible, relative to the kinetic energy of the molecules at a given temperature.

Numerous experiments have shown that exposure to electromagnetic fields of super-low intensity (below kT) causes detectable effects in water systems. The system of extremely high-frequency therapy using low-intensity electromagnetic radiation (in the millimeter band†) is built upon these principles. Long-term clinical trials have proven to be effective.

In the previous chapters, we discussed the work of groups

under the leadership of E. B. Burlakova and A.I. Konovalov, who studied the effects of ultra-low doses of various substances upon biological objects. Similar effects were detected when studying the impact of ultra-low doses of ionizing and electromagnetic radiation on living organisms. A bimodal dependence of parameters is shown from the dose: the effect increased at low doses, reaching a maximum at a particularly low dose, then with increasing the doses, the effect decreased (in some cases changing the sign of the effect) and then increased again with the next increase of the dose. The value of the low dose's maximum effect and the dose, at which it occurred, depended on the nature of the biological objects and dose's intensity. The findings indicate that the reaction of the organism to the action of low doses of radiation is a function of the intensity of radiation plus the time elapsed since the beginning of the irradiation. When there is a biphasic response to an exposure to increasing amounts of a substance or condition, this process is described as hormesis.

The results show that water, which contains ions of $Ca_2+$, $Na+$, $K+$ and $Cl-$ treated by a weak combined constant (42 mcT) and a low-frequency (0.06 mcT) magnetic field, caused the fluorescence of bovine serum albumin. The magnitude of the effects depended on the frequency of the alternating field and the combination of ions. After treatment, the solution contained fairly large (700-900 daltons) and stable molecular associates.[42]

Electromagnetic fields (EMFs) of low-intensity (power flux density $3.10^{-4}$ watts.cm-2 at a wavelength of 3.2 cm) create a temperature gradient sufficient to alter the gas exchange (coalescence of bubbles), which reduces the concentration of the air dissolved in the extra-cellular medium. The influence of the weak combined constant (42 $\mu T$) and alternating low frequency (40 nT; 3-5 Hz) magnetic fields changes the intensity of self-fluorescence of a number of enzymes.

Correlations between changes in the functional activity of enzymes and the possible transmission of the effect through treated magnetic field solvent (water, 0.01 M NaCl) were shown[43].

[42] Ayrapetyan S.N., Grigorian K.V., Avanesian A.S., Stamboltsian K.V. Magnetic fields alter electrical properties of solutions and their physiological effects. Bioelectromagnetics.;15(2):133–42. 1994

[43] Oshitani J., Uehara R., Higashitani K. Magnetic Effects on Electrolyte

Magnetic field parameters were close to the characteristics of the geomagnetic field, initiating the processes of chemical reactivity and ionic conductivity in aqueous solutions of organic molecules. The assumption is, that this effect can be explained by the cooperative interaction of the ensemble of a large number of ions with the weak fields.

The infradian rhythm of behavioral reactions, along with the temperature and body mass of epifizarnah rats exposed to EMF (at a frequency of 8 Hz) and induction 5 mcT were studied in a 40-day experiment. The researchers a discovered a restructuring of the spectrum of rhythms, an increase of their amplitude, and the development of desynchronosis. There was also a significant change in the content of free oxygen in water and aqueous electrolyte solutions under the action of the electromagnetic radiation within the millimeter range. The conclusion: the water component of the solutions could be a primary target in the action of extremely high-frequency EMFs on biological objects.

We emphasize, again, that when studying the actions of weak electromagnetic fields upon biological systems, we define the same set of properties as in the study of ionizing radiation and the chemical action of biologically active compounds. However, the influence of EMF has its own specifics. For example, the bimodality dose dependence is transformed in the presence of the so-called amplitude and frequency windows. There are intervals of frequencies (or even single-frequency resonances) and intervals of amplitudes, at which the effects are recorded. Outside of these windows, the effect may be absent. Frequency, as an independent parameter, may, in a sense play the role of the dose. When the frequency changes, so too, does the magnitude of the effect vary and can even change sign.

Another general property of ultra weak influences is their effectiveness at higher levels of exogenous EMF. This has been repeatedly demonstrated at different levels of organization. At the level of the organism, for example, the effectiveness of EMF was found at the level of induction of 1-5 nT (for the geomagnetic field, this value varies from units of ten to even hundreds of nT) and an 8 Hz frequency—but with prolonged exposure.

Primarily sensitive were the cardiovascular system, blood

system and nervous systems. It's possible to develop a person's conditioned response at the frequency of 0.7 MHz and EMF amplitude $10^{-4}$ V/m.

Physical agents, capable of ultra-low levels of exposure, significantly change the state of the biological systems and are not limited to electromagnetic fields alone. There is evidence of high biological activity for infra-low-frequency acoustic fields. It's possible the direct biological effect has background neutrons, etc.

Negative air ions play a special role. The history of their study goes back to the beginning of XXth century. In the 1920s, studies showed that a vanishingly small concentration of negative ions in the air is absolutely crucial for the normal functioning of organisms.

Problems associated with a manned flight to Mars were studied in a joint USA/USSR project. One such problem to resolve concerned the influence of prolonged radiation exposure on the body during flight. They studied the effect of low-intensity radiation (gamma Co60) in a dose of 20-25 cGr for the duration of a year on the metabolism of irradiated dogs. These experiments lasted from three to six years.

The study determined the main damage under neutron irradiation to be spermatogenesis. Therefore, the main conclusion proposed was that only people beyond childbearing age need to be considered to join a Mars expedition. Also, one of the important consequences of low-intensity exposure was the fact that irradiated dogs changed their tolerance and sensitivity to any additional factors introduced, both chemical and physical in nature.

For example, despite the fact that their blood picture remained practically unchanged during the irradiation, the influence of drugs that caused leukopenia, led to serious violations of leucopoiesis in irradiated dogs (who received much lower doses than in the controls). A similar situation was observed when irradiated dogs were subjected to physical loads (e.g., running on a treadmill): dogs either refused to run on a treadmill, or were unable to finish the distance, although 100% of the control had finished.

The review by Voeikov V. L., in 1998, cited various examples of the reactions of living organisms to low-intensity factors. They can operate as information signals in natural conditions in examples of directional exposure to low-intensive factors of an artificial nature. Voeikov discusses the reasons why, until recently, mainstream biology rejected the possibility of such influences. He discusses, as well, the theoretical ideas of E. Bauer regarding the structural and energetic specificity of living systems.

Bauer formulated "the principles of sustainable disequilibrium." According to these principles, all living systems are constantly in a high state of non-equilibrium and use their energy resources to maintain this state. This explains their high sensitivity to the action of low-intensity factors. This is a basic property of living systems, which are traced on all levels of their organization—from the molecular to the cellular. In particular, as shown in the survey data, a portion of the macromolecular components of a living cell energetically transform into electronically excited states, which follows from the ability of living systems to emit photons in a wide range of the electromagnetic spectrum, including ultraviolet.

So, biological systems have a very high sensitivity to weak and super-weak physical influences—in particular, to fluctuations in the audio frequency range, which presumably have an impact on them through their aquatic environment. Perhaps the effects are due to the fact that both cytoplasmic and the intercellular water medium of living organisms have a very complex organization. This organization is different from ordinary water and displays the properties of a "microphone."

Indeed, the water in living cells and in dead cells differ. So, if you destroy living cells to "homogenate" then centrifuge it at very a high rpm, the contents of the cell would split into two fractions: water-insoluble and water-soluble. In the sludge, will be the water-insoluble components from the cell's membrane—organelles surrounded by membranes, ribosomes, etc... The supernatant liquid would contain the so-called water-soluble components of various proteins and peptides along with free RNA, and other macromolecules—not to mention the low molecular weight of any organic molecules present within the cells.

At first glance, the same thing happens to large living cells (a single-celled Alga Euglena, for instance), when centrifuged even at low speeds. Large particles sink down nuclei, mitochondria, etc., while fat droplets float to the top, and are separated by the aqueous phase. But, surprisingly, in this phase, there are no enzymes, proteins, or nucleic acids, which can be considered "water-soluble." Moreover, the experiments of A. G. Gurvich in the 1930s showed that even after centrifugation, the parts retained the ability of mitosis. Hence, the components of living cells are "water soluble" only after the cell's destruction. In other words, the waters in the homogenate and in the living cell differ from each other significantly.

The results obtained in experiments on the direct measurement of the speed of water diffusion in living cells indicate a distinction within these "waters." According to the data supplied by many authors, only from one quarter to one third of the cell water has the same movement speed as that of "normal" water. The rest of it has some other mobility, we can describe as "structured." The amount of collaborating data is increasing, and this causes us to revise many of our preconceived ideas about the organization of the cell cytoplasm.

It turns out that cytoplasm is not a solution where components randomly interact and collide with each other. Cytoplasm can be likened more to jelly, which begins to "tremble" in response to external stimuli. But this comparison is all very relative, because the cytoplasm is penetrated by numerous "pores" which are organized by the flow of metabolites to the sites of their processing. Due to this structure, the cell works as a single unit because signals from one part are immediately transferred to all other parts.

A pronounced therapeutic effect of low intensity microwave radiation (frequency range 30-300 GHz, lengths of EM-waves 1-10 mm) explains the special properties of water in the body and by the fact that a significant part of it is in the bound state. A thorough study of the influence of low-intensive mm-radiation on "normal" water and aqueous solutions showed that the irradiation of solutions being in equilibrium did not lead to any registered changes of their physical parameters.

But, it turned out, that it is still possible to find such conditions—when low-intensive mm-radiation affects the properties of water systems outside of living organisms. This occurs when irradiation is applied to water systems in a state of non-equilibrium: like water-salt systems with semi-permeable walls, dispersed water, supersaturated solutions of gases and salts, water containing cavitation bubbles and radicals. Under the action of the weak pulses of the microwave field, within such systems of non-equilibrium may appear additional zones of non-equilibrium with a long relaxation time. This may be attributed to water "memory."

Pure physical experiments can detect the "memory" of water. For many decades, the subject of how the property of water changes under the influence of magnetic fields was hotly debated. Some water experts still categorically deny the very possibility that water can "remember" the influence of magnetic fields. On the other hand, magnetic fields are already used in water treatment techniques. Since magnetic fields make water more "smooth," the economic efficiency of water projects is dramatically increased.

In fact, with "normal" technical water and with water treated by a constant magnetic field, carbonate crystals differ in their properties. During the magnetic treatment of water salts (primarily calcium carbonates), it is not the creation of rigidity, which causes the salts to disappear. The danger for pipes and boilers is calcite—a very solid crystal, tightly adhering to the walls. After magnetic treatment of water, calcium carbonate precipitates as aragonite— small soft crystals that are driven with a water currents, which can easily be removed through filtration. Moreover, such water can loosen any calcite already deposited on the walls, by converting it to aragonite, which is carried away with the water flow.

We assume that the nature of the crystals formed in the water depends upon the degree of the water's non-equilibrium. In active (living) water, the connection between the elements of the crystals is weak. In non-activated water, the connections are strong, leading to the formation of solid crystals. This kind of consideration can easily be applied to the formation of stones in the human body that is, low energy contributes to stone formation.

The Fluid Magnetic Corp., in Dinuba, CA, USA conducted detailed research into the effect of magnetic fields upon water. They found that their magnets can affect only water that is running—not standing water. The magnets' effectiveness greatly depends on the flow rate of water, the diameter of the pipe, the intensity of the magnetic field, the direction of the magnetic force lines, temperature, pH, and other more subtle properties of water. In this case, as in all the others described above, the effect depends not so much on the "dose" of the magnetic field, but on certain combinations of dynamic parameters—parameters determined by the degree of the non-equilibrium of the water being treated. If optimal parameters for the magnetization are chosen, the water "memory," (recorded by the type of produced by its evaporation crystals of $CaCO_3$,) will persist for more than 2 days after contact with the magnetic field.

The science of the nineteenth century was devoted to the study of ideal systems in the state of rest, or in a state of simple movement. The science of the twentieth century significantly expanded the horizon of understanding the world. A qualitative transition to the probabilistic description of dynamic processes was made. By and large, this description has been reduced to snapshots of the processes. The best case takes into account the simultaneous interaction of two or three reagents. This allowed elegant models of the physical and chemical processes to be created, but these models were fleeting, operating in an instant.

Like any model, these were limited, describing only certain aspects of reality. But as is always the case in the history of science, these imprecise models were accepted as an adequate reflection of objective reality. The prevailing assumption back then, was that the modern scientific paradigm fully describes the world around us. Thus, trying to understand any complexity, dynamism, multi-component nature of the processes, virtually disappeared from consideration. The world was described as linear, logical, and predictable—where every action caused a known reaction.

This paradigm is the basic principle of modern classical medicine. The vision of this paradigm insists: certain medications cause known reactions; all processes in the organism are based on chemical interactions; by controlling these reactions, we can predictably affect the state. Genes dictate the characteristics of the organism, and permutation allows you to manage past, present and the future. We understand processes at the molecular level and can manage these processes. Surgeons could swap new body parts for worn-out organs so their patients can begin a new cycle of life. For a little more money, a new body can simply be grown in an autoclave and provided to a paying client.

However, such representations turn out to be just a big illusion. A hundred years of development of Western medicine, despite the enormous efforts and resources invested in this sector, show that it is still impossible to cope with many major diseases. Instead of only infectious and acute diseases, we currently experience cardiovascular, cancer and chronic diseases. Too bad all the talk about cloning and genetic modification turned out to be just hype for the fraudulent purposes of deceiving investors for their money. The big problem remains—all modern biological concepts are still based on the faulty assertion that we live in a linear, logical world.

Simple linear thinking has been employed in the study of water and water systems where chemical formulas, the basic properties and the composition of impurities, both useful and harmful to life, have all been defined. Yet, at some point, a debate started: why do we need minerals in water? Isn't it better to drink super-purified—distilled water? (Let's not even think about the source from which it was pumped out.)

This view was adopted in a number of countries. But, as it turns out, the distillate is not so useful. It leaches out minerals in the body, making bones and hair more brittle. So they added salt to the distillate, declaring the resulting water to be beneficial for life. Still, there was something lacking. Natural water is still better. But how can this be proved? Where is the authoritative word of science?

Apparently, in this case, we again face a familiar situation. Nature is not as simple and predictable as we would like to believe, especially when it comes to human health and, in particular, food. I remember the enthusiasm with which it was perceived the invention of hydroponics—growing vegetables and fruits in aqueous solutions. The usual problems of gardening did not exist, like the need for fertilizer and the presence of bugs and other pests.

In closed factories, under artificial sun, rows of plastic cylinders are settled, in which a solution is pumped. In an ideal environment for growing, big tomatoes and red round apples ripen. Here it is—a triumph of civilization over nature! But soon it became apparent that the tomatoes and apples had no smell or taste—more like plastic toys than food. Smell the fruits and vegetables from the market and you feel the difference.

The same story is with water. Make a simple experiment: water houseplants with natural and artificial water. After a week, you will see a difference. Fish are even more sensitive to the composition of water—a very well known fact to any one who keeps an aquarium. So we need to employ scientific research methods. And here, it turns out, the investigation of water is not such a simple task.

One of the basic corner stones of classical physics is that the object undergoing study always remains the same under similar conditions—so that different laboratories at different times of day and the year can obtain the same results. This is the basis of the principles of classical physics and chemistry. However, as soon as we begin exploring water, we find that its properties change in unpredictable ways. Naturally, we are usually talking about extremely small variations of the basic parameters, but it turns out that these variations are not random. And in some cases, the variation of parameters in the original homogeneous aqueous systems may eventually become very large.

Let's take a hermetically sealed bottle of distilled water, open it, then pour the water into several receptacles, glass and plastic. Then we measure the basic parameters, for: pH, specific conductivity, amount of dissolved oxygen, the total concentration of electrolytes, and the super-weak luminescence of the liquid induced by the electric field (electro-photonic) or chemically (bio-photons). We leave these vessels sit for a day, and repeat the measurements. We make sure that the parameters in the different vessels would be different.

This simple experiment shows that water parameters depend on a number of factors, and vary dynamically in time. It is possible to identify the main factors influencing water parameters:

- the interaction with the gases of the air;
- the uptake of dust from the air;
- exposure to light, especially sunlight;
- the interaction of water with the vessel walls, especially of plastic;
- the formation of water layers around the molecules of impurities;
- the secretion of gases from the water—mainly oxygen and hydrogen.

Of course, these processes are highly dependent on the temperature and pressure of the surrounding air. But there are also internal factors related to these non-linear dynamic processes occurring in water, in particular, its auto-oxidation and self-restoration.

To minimize the impact of all these factors, the researchers usually practiced strict controls over all the conditions, and most importantly, used bi-distillate water for their experiments, which is free from any impurities. The results, of course, have great scientific value, but they do not address anything about the properties of water that surround us in daily life. This is because water contains various impurities that radically change its behavior.

We can assume that the coherent domains (CD) are likely to be the main place for the development of complex and fast biological reactions. It is happening due to the ability to form mixed CD by sets of amino acids, peptides, proteins and other biologically important molecules. Given the high-energy bonds between the molecules that form the CD, we can assume that mixed CD may serve as links between molecules, spectrally compatible with water molecules.

Therefore, it is a CD, and not the weak hydrogen dipole-electrostatic bonds, which can be destroyed easily by thermal noise. The CD could provide a strong connection of diverse bio-molecules or different areas of the same large molecule into a single unit, for example, in the formation of fairly stable tertiary and quaternary structures of protein.

The flow velocity of the enzymatic complex biochemical reactions in the organisms of mammals was a long-standing unsolved problem in molecular biology. Being within the domain of coherence, water has a viscosity more than an order of magnitude smaller than in the incoherent environment. As a result, due to the remarkable properties of the domains, all chemical reactions occur in them an order of magnitude faster than in the incoherent environment.

Quantum electrodynamics' description of aqueous solutions has solved some problems of molecular biology. First of all, it refers to a complex multi-vital enzymatic reaction, the rate of which would be clearly insufficient for the human body, in case the coherent domains were absent in the biological fluids of living organisms.

# Water as receiver and transmitter of information

*Of course, we all know that water is transparent and has no taste, smell, or color. What's amazing about water is that it is able to show us myriad "faces" depending upon the condition it is in or the information it is given.*

Masaru Emoto (1943-2014)

Luc Montagnier, virologist, won the Nobel Prize in medicine in 2008 for his discovery of the human immunodeficiency virus (HIV). He was a Professor at the Pasteur Institute in Paris and Director of Research at the French National Center for Scientific Research (CNRS). After 2008, he continued the work of Benveniste, completely changing the direction of scientific research. This has led to some surprising results that still cause fierce opposition in the fore. To present the main results of Montaignier's inquiries in this direction, we give the floor to him:

Luc Montagnier's story begins more than ten years ago, when he began to study the strange behavior of a small bacterium. Mycoplasma pirum (M. pirum) is a frequent companion of HIV, and like HIV, is a lover of human lymphocytes. Using filters of 100 nm and 20 nm, Montagnier tried to separate the bacterium—which are about 300 nm in size—from viral particles whose size, by filtration, is about 120 nm. Starting with pure cultures of the bacterium on lymphocytes, the filtrates were indeed sterile for the bacterium, when cultured on the rich cellular medium, SP4.

Polymerase chain reaction (PCR) and nested PCR adhesion—based on primers derived from a gene of M. pirum, which had previously been cloned and sequenced—was negative in the filtrate. However, when the filtrate was incubated with human lymphocytes, (previously controlled for not being infected with the mycoplasma) the mycoplasma, with all its characteristics, was regularly recovered! Then the question was raised: what kind of information could be transmitted in the aqueous filtrate? This became the beginning point of a long-lasting investigation bearing on the physical properties of DNA in water. Indeed, a new property of M. pirum DNA was discovered: that the emission of low frequency waves in some water dilutions of the filtrate soon extended to other bacterial and viral DNAs.

A solenoid comprises the apparatus used to detect electromagnetic signals and captures the magnetic component of the waves produced by the DNA solution in a plastic tube converting the signals into an electric current. These currents are then amplified and finally analyzed in a laptop computer using specific software (Figure 16).

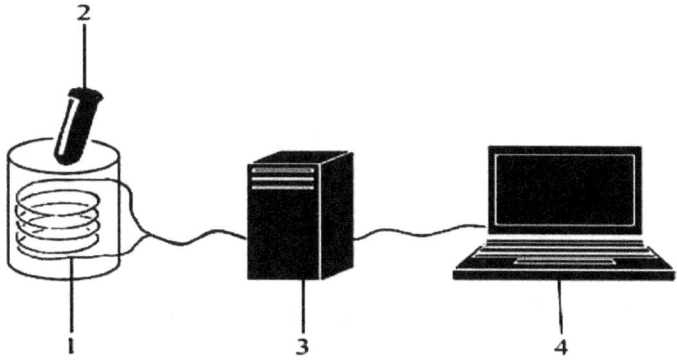

*Fig. 16. This device is used for the capture and analysis of em signals. (1) Coil made up of copper wire, impedance 300 Ohms. (2) Tube with plastic stopper containing 1 ml of the solution to be analyzed. (3) Amplifier. (4) Computer.*

Following is a brief summary of the laboratory observations:

1) Ultra Low Frequency Electromagnetic Waves (ULF 500–3000 Hz) were detected in certain dilutions of filtrates (100 nm, 20 nm) from cultures of microorganisms (virus, bacteria) or from the plasma of humans infected with the same agents (Figure 17). The same results were obtained from their extracted DNA.

2) The electromagnetic signals (EMS) are not linearly correlated with the initial number of bacterial cells before their filtration. In

one experiment, the EMS became similar in suspension to the E. coli cells, varying from 109 down to 10. It's an all or nothing phenomenon.

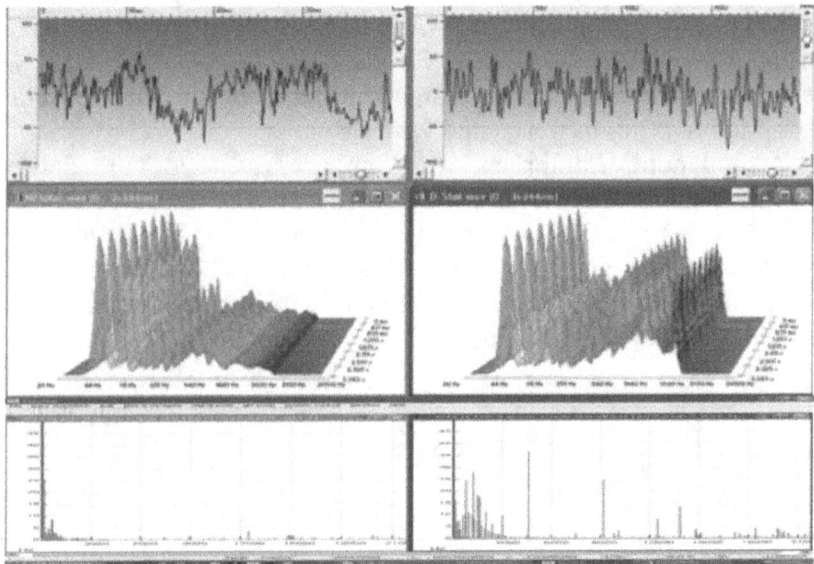

*Fig. 17. A graph showing typical signals from aqueous dilutions of M. pirum* (Matlab software). *Note the positive signals from D-7 to D-12 dilutions.*

3) EMS is observed only in some high water dilutions of the filtrates. For example, from 10–9 to 10–18 dilutions are shown in some preparations of E. coli filtrates.

4) Some bacteria don't produce EMS: this is the case of probiotic bacteria, such as Lactobacillus and also of some laboratory strains of E. coli used as a cloning vector.

5) These studies have been extended to viruses, although not all the viral families have been explored. Similar EMS was detected from some exogenous retroviruses (HIV, FeLV), hepatitis viruses (HBV, HCV), and influenza A (in vitro cultures). The technical conditions for EMS induction is summarized by the following:

**Use high dilutions in water—mechanically agitate (Vortex) between each dilution.**

Excitation by the electromagnetic background of extremely low frequency (ELF), starts very low, at 7 Hz. The excitation is not induced when a mu-metal cage shields the system. The stimulation by the electromagnetic background of a very low frequency is essential. The background is either produced from natural sources (the Schumann resonances which start at 7.83 Hz) or from artificial sources.

## A DNA sequence transmitted through waves and water.

In further experiments, a fragment of HIV DNA, taken from its long terminal repeat (LTR), has been used as the DNA source. This fragment was amplified by PCR (487 base pairs) and nested PCR (104 base pairs) using specific primers. In a first step, DNA dilutions were made, in which the production of EMS under the ambient electromagnetic background was detected.

Then the following steps were taken. As shown in Figure 3, one of the positive dilutions (say $10^{-6}$) was placed in a container shielded by 1 mm thick layer of mu-metal (an alloy absorbing ultra-low frequency waves). In its vicinity, another tube containing pure water was placed. The water content of each tube was filtered through 450 nm and 20 nm filters and diluted from $10^{-2}$ to $10^{-15}$.

The copper solenoid placed around them, receives a low intensity electric current produced by an external generator oscillating at 7 Hz. The produced magnetic field was maintained for 18 hours at room temperature. Then, EMS was recorded from each tube.

Now, also, the tube containing the water emits EMS at the dilutions corresponding to those positive for EMS in the original DNA tube. This result shows that, when achieving a 7 Hz excitation, the transmission into pure water of the oscillation of the nanostructures (initially originated from DNA) has been achieved. The following controls were found to suppress the EMS transmission in the water tube:

• Time of exposure of the two tubes (less than 16 – 18 hrs) - No coil

• Generator of magnetic field turned off

• Frequency of excitation < 7 Hz - Absence of DNA in tube 1.

At this point, the most critical step was undertaken—namely to investigate the specificity of the induced water nanostructures by recreating the DNA sequence from them. For this, all the ingredients to synthesize DNA by polymerase chain reactions (nucleotides, primers, polymerase) were added to the tube of signalized water. The amplification was performed in a thermo-cycler under classical conditions (35 cycles). The DNA produced was then submitted to electrophoresis in an agarose gel.

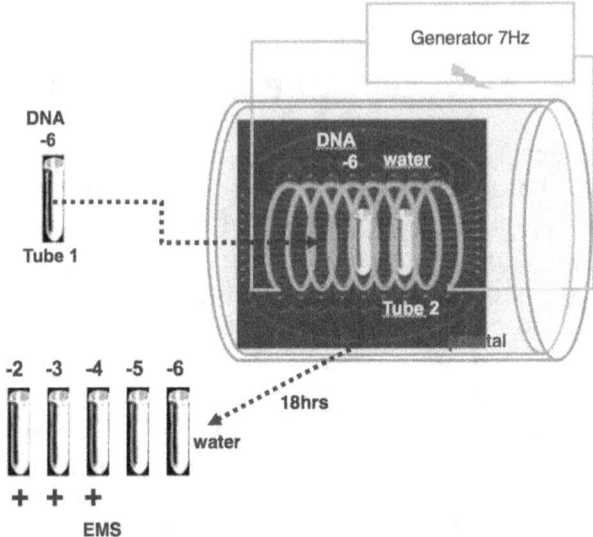

*Fig. 18. A diagram for the transmission of DNA genetic information into water through electromagnetic waves.*

The result was that a DNA band of the expected size of the original LTR fragment was detected. Further verified was the fact that this DNA had a sequence identical (or close to identical) to the original DNA sequence of the LTR. In fact, it was 98% identical (only a 2 nucleotide difference) out of 104. This experiment was found to be highly reproducible (12 out of 12) and was also repeated with another DNA sequence from the bacterium, Borrelia burgdorferi, the agent of Lyme disease.

It became clear that the water nanostructures and their electromagnetic resonance could faithfully perpetuate DNA information. These elements give support to a provocative explanation of our Mycoplasma Pirum Filtration experiment. The nanostructures induced by M. pirum DNA in the filtered water represent different segments of its genomic DNA. Each nanostructure, when in contact with the human lymphocytes, is retro-transcribed in the corresponding DNA by some cellular DNA polymerases.

Then, there is a certain probability (even very low) that each piece of DNA recombines within the same cell to other pieces for reconstructing the whole DNA genome. We have to assume that, in presence of the eukaryotic cells, the synthesis of the mycoplasma components (membrane lipids, ribosomes) can be also instructed by the mycoplasma DNA. One single complete mycoplasma cell is then sufficient to generate the whole infection of lymphocytes.

Recent experiments of the G. Vinter group have shown that a synthetic genomic DNA is sufficient to maintain all the characteristics of a mycoplasma. All the steps assumed in the regeneration from water can be analyzed and open to verification.

## The Theoretical Framework

The above experimental observations fit into the physical view, which addresses biological dynamics as an interplay between chemical processes and em interactions; i.e., as an array of em-assisted biochemical reactions. The above experimental results were interpreted within the framework of a theory of liquid water based on the Quantum Field Theory developed by Emilio Del Giuidice. Montagnier and Del Giuidice published several articles together. This theory is intrinsically non-linear. It provides tools suitable to describe a complex ensemble of processes, which are also non-linear.

In accordance with this theory, an electromagnetic signal is generated by the rotation of the plasma of quasi-free electrons of the coherent domains. DNA and the vast majority of proteins are poly-anions, so are structured water, and the cloud of positively charged ions—some of which have a cyclotron resonance frequency in the range 1100 Hz surrounding them. Thus, when a magnetic field is applied, whose frequency corresponds to the frequency of cyclotron resonance of ions, these ions are shifted from their orbits. In particular, the role played by low em frequency in the background is understood by observing, that we need a resonant alternating magnetic field in order to load energy in the water, CDs.

In higher organisms, such as the humans, this field can be produced by the nervous system. Elementary organisms, such as bacteria, use environmental fields. Good candidates are the Schumann modes of the geomagnetic field. These modes are the stationary modes produced by the magnetic activity (lightning or else) occurring in the shell whose boundaries are the surface of the Earth and the conductive ionosphere, which acts as a mirror wall for the wavelengths higher than several hundreds of meters.

Based on all this data, Montagnier proposed a number of ideas around the electromagnetic nature of various diseases. The list of diseases in which EMS have been found (Alzheimers, Parkinsons, Multiple Sclerosis, various neuropathies, chronic Lyme syndrome, Rheumatoid Arthritis), indicate clearly that their presence is not limited to diseases known to be of infectious origin. The fact that EMS have been found in diseases not known to be of infectious origin is intriguing, and leads us to seek bacterial or viral factors in these diseases.

Later on, they developed a method of photonic transmission of the DNA signal to the distant laboratories. An EMS signal could be successfully detected, digitized, send by Internet, transformed into analog form, then applied to water (with electromagnetic field). After using PCR reaction for some time, it was possible to detect initial DNA from pure water. This experiment was performed successfully from France to Italy.

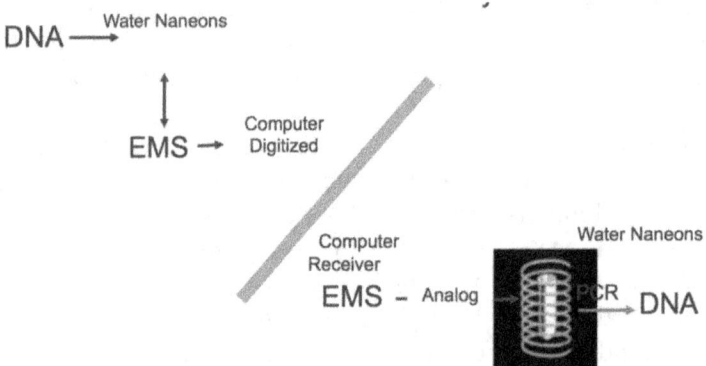

Fig. 19. *The principle of DNA signal transmission.*

As we can see from this brief overview, many of the results and ideas of Montagnier correlate with the data of Konovalov. It would be important to look at the presence of the electromagnetic signal in Konovalov's solutions. A lot of work is ahead for a whole generation of researchers. Without the success of multiple reproductions receiving the same outcome, the results would have continued to be regarded with caution. Let's hope that the next generation of young scholars will take up the baton.

# Water as a Source of Light

*I Gathered the Aura of Light Around Me*
*The Aura of Light, Softly Surrounds Me*
*An Aura of Light, So Bright*
*I Closed Eyes to View Sight...*
MoonBee Canady, modern-time poet

To study structured water, we need different experimental techniques. One of these is the Electrophonic Imaging (EPI) method, also known as the Gas Discharge Visualization (GDV) technique. The study of Electro-photonic parameters of liquids is based on using a commercially produced instrument called the "Bio-Well," (www.bio-well.com). This instrument is well-known for analyzing stimulated photon emissions from human fingers, which is used for health and well-being diagnostics; the analysis of athletes; of altered states of consciousness; of the influence of music and yoga on people; as well as of geo-active zones and minerals[44].

When the EPI parameters are measured for liquid subjects, a drop of the liquid is suspended at 2-3 mm distance above the glass surface of the optical window of the device, and the glow from the meniscus of the liquid is registered (Figure 20). The volume of the liquid is about $5*10^{-3}$ ml. The temperature is kept in the range of 22-24 C; the relative humidity is maintained from 42% to 44%. The train of triangular bipolar electrical 10 mcs impulses of amplitude (3 kV at a steep rate of $10^6$ V/s and a repetition frequency of 103 Hz), is applied to the conductive transparent layer at the back side of the quartz electrode, thus generating an electromagnetic field (EMF) at the surface of the electrode and around the drop.

Under the influence of this field, the drop produces a burst of electron-ion emissions and optical radiation light quanta in the visual and ultraviolet light regions of the electromagnetic spectrum. These particles and ions initiate electron-ion avalanches, which give rise to the sliding gas discharge along the dielectric surface. A spatial distribution of discharge channels is registered through a glass electrode by the optical system with a charge-

---

[44] Korotkov K. G. The Energy of Health. Amazon.com Publishing, 2017

coupled device TV camera, before being digitized in the computer.

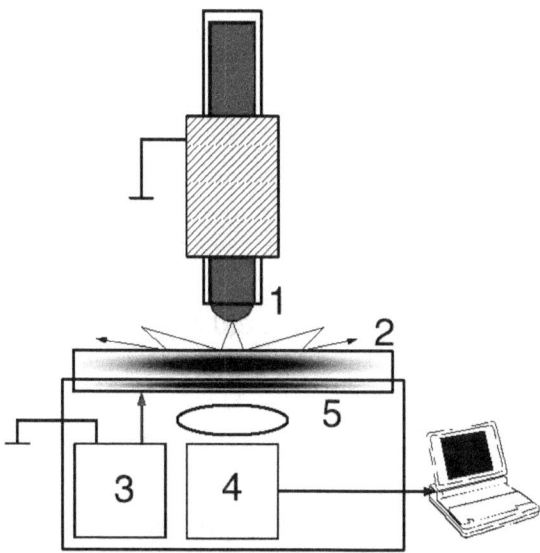

*Fig.20. The principle method behind the study of the electro-photonic glow of liquids. 1 – liquid meniscus; 2 – transparent quartz electrode; 3 – impulse generator; 4 – optical system; 5 – conductive coating.*

The drops are exposed to the EMF from 2s up to 10s, and short "films" are recorded in the computer as .avi files. The avi files are then converted to a series of BMP files, and the area (the number of light-struck pixels) and averaged intensity (ranked from 0 for absolute black to 255 for absolute white) parameters for the software calculates every image. The time series are averaged on 10 measurements that provide the statistical reliability at the confidence level of 0.95, with the experimental sensitivity of 75%. Examples of the EPI glow for different liquids are presented in Figure 21.

This method is very sensitive to the condition of water and liquids. The sensitivity of the Electro-photonic analysis of water allows for the study of the changes in the properties of water under different influences. As an example Figure 3 demonstrates an EPI signal from several samples of tap and filtered water under the influence of a pulsating magnetic field. In all cases, we see significant changes of the EPI parameters of the water samples[45].

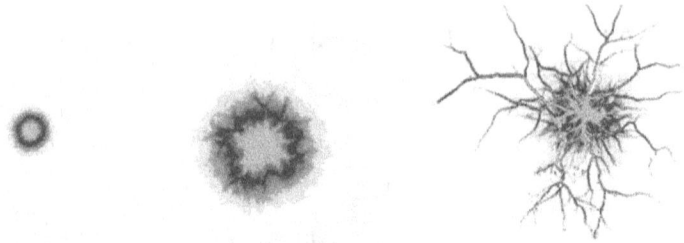

*Fig. 21. Different examples of the EPI glow from different samples of water: 1 – distilled water; 2 – tap water; 3 – structured water.*

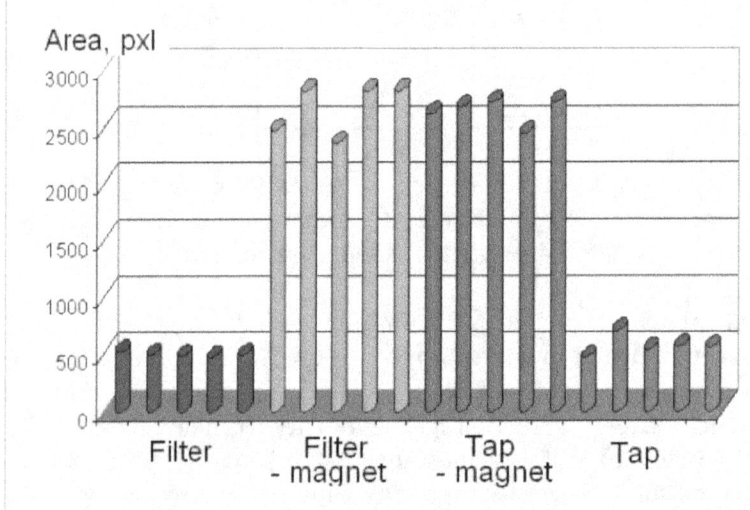

*Fig.22. EPI Area for initial and magnetized samples of water.*

The study of water should be undertaken with great precautions: water changes its properties in the process of interaction with air after opening the bottle. This process is well known in the wine industry. We may see the same type of "aging" is also true of water. As an example, Figure 23 demonstrates the change of time dynamics of electro-photonic parameters for two water samples: the bottle, just after opening and 4 hours later. The

---

[45] Korotkov K.G. et.al. The Research of the Time Dynamics of Gas Discharge Glow around Drops of Liquids. J of Applied Physics, 2004

water was taken from two commercially available plastic bottles of "Evian" water purchased in the same shop.

The bottles were opened simultaneously, in the same laboratory room, at room conditions. The results showed that immediately after opening the bottle, the water's glow was characterized by a great variability between the measurements and by a considerable increase of the values of parameters, in two distinct phases. But about 2 minutes after opening, the water stabilized. For the samples of water that were taken 4 hours later, a rise can be observed during the first 30 seconds. After that, the parameters remained stable. The amplitude of glow for the distilled water is considerably lower and practically remains unchanged with the passage of time.

To study the dependence of EPI parameters on the concentration of salt solutions of strong electrolytes (completely dissociated into ions by solvents such as NaC1, KC1, NaNO$_3$) in distilled water were used. The EPI parameters reach a quasi-stable level at some concentration, which is related to a particular electro-conductivity of the solution.

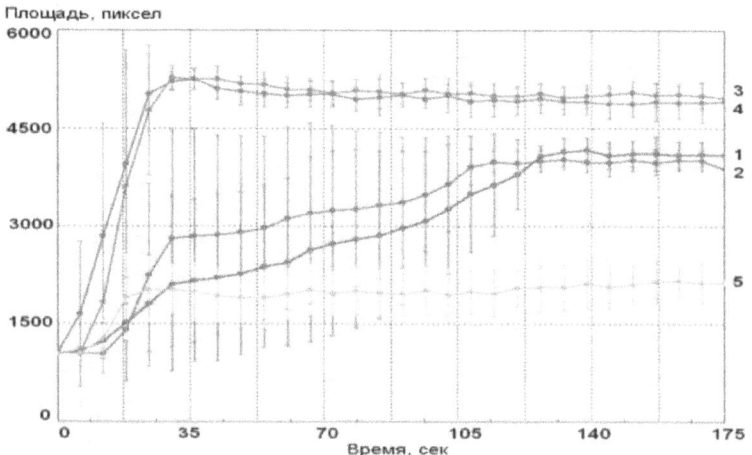

*Fig. 23. The time dependence of the EPI glow area of the water drops. 1,2 – after opening the bottle; 3,4 – 2 hours after opening the bottle; 5 – distilled water.*

The studied solutions of strong electrolytes are different in ion radiuses and electro-conductivity. They have statistically significant differences between EPI parameters for neighboring concentrations of the same solution as well as between similar concentrations of different solutions. The glow area of drops of the same size is a function of photon emissions from their surface. Consequently, activity and flexibility of ions, as well as the degree of ionization and dissociation, will contribute to the value of the given parameter. The Area parameter and the equivalent electro-conductivity appear to be connected by the polynomial dependency of the fifth range.

Different researchers are using the EPI method. A group from Arizona University conducted research using GDV/EPI technology to differentiate ultra-molecular doses of homeopathic remedies from solvent controls and from each other[46]. It was blinded, randomized assessment of four split samples each of 30C potencies of three homeopathic remedies from different kingdoms—for example, Natrum muriaticum (mineral), Pulsatilla (plant), and Lachesis (animal), dissolved in a 20% alcohol-water solvent versus two different control solutions (that is, solvent with untreated lactose/sucrose pellets and unsuccussed solvent alone). GDV measurements, involving application of a brief electrical impulse at four different voltage levels, were performed over 10 successive images on each of 10 drops from each bottle (total 400 images per test solution per voltage). A very detailed research project showed that homeopathic remedies of 30C potency had different responses to the electromagnetic field, as compared with the solvent. Their EPI parameters were statistically different as well. The findings suggest that the biophysical method of GDV may allow differentiation of ultra-molecular doses of homeopathic remedies from solvent controls and perhaps from each other at specific doses of voltage amplitudes under blinded conditions.

Researchers concluded: "If standardized, the GDV testing in the clinical setting might eventually allow differentiation of active from inactive remedy supplies at the time of administration to patients. The ability of GDV to assess both liquids and human subjects (testing fingers or toes) could facilitate studies in which both the individually chosen remedy and the patient undergo testing at

---

[46] Bell I., Lewis D.A., Brooks A.J., Lewis S.E., Schwartz G.E. Gas Discharge Visualisation Evaluation of Ultramolecular Doses of Homeopathic Medicines Under Blinded, Controlled Conditions. J of Alternative and Complementary Medicine, 2003, 9, 1,. 25–37.

baseline and follow-up to determine any objective, measurable characteristics that might predict better or worse clinical outcomes."

Similar research was done with the Bach Flower essences. They were prepared by adding several drops of the flower essence to the solvent. Figure 7 demonstrates the results of the experiment, where a different number of drops of chamomile essential oil was added to 100 ml of solvent. As we see from the graph, the highest EPI signal was found from solvents with half a drop and two drops. How interesting it is, that in Dr. Bach's recommendations, he wrote to just add "two drops."

A group from the Estee Lauder Company prepared aqueous gemstone elixirs by crushing gemstones into a fine powder and dissolving them in DI water (0.5% w/v)[47]. After 10 minutes of stirring with a magnetic stirring bar at room temperature, the elixir suspension was filtered through coarse P8 filter paper to remove gemstones from the elixir. In the citrine control, the citrine gemstone suspension was placed on the magnetic stirrer without a stirring bar. This experiment was designed to test the effect of stirring in the presence of a magnetic field, which was measured to be 250 T. In some cases the gemstones were left undisturbed in the elixir suspension for 1-7 days at room temperature. Immediately before electro-chemical and GDV measures, all elixir suspensions were filtered through a 0.2 m Millipore filter. In some cases, the elixirs were mechanically agitated or succussed immediately before the second filtration. Manually shaking and banging the elixir on a hard surface three times accomplished succussion.

The results demonstrated increases in GDV area for all elixirs when compared to the solvent control (Figure 24). These results are consistent with the conclusion that minerals are leached from the gemstones and become soluble in the aqueous elixirs. The next series of experiments were designed to measure the differences between elixirs with the gemstones removed immediately after preparation and those where the gemstones remained.

They predicted that the amount of leached mineral would increase over the course of several days. The results of this experiment demonstrated an increase of GDV Area during first 3 days. At the same time no statistically significant differences were

---

[47] Vainshelboim A.L., Hayes M.T., Momoh K.S. Aveda advertisement – Tourmaline Charged Radiance Fluid Jane Magazine. August 2004. pp. 24-25

found between elixirs with gemstones left in or removed. Similar values between these samples were obtained both for conductivity and GDV area measures. This suggests that the initial leaching of minerals from the gemstones is the primary process of concern. After four days in the presence of gemstones, no additional minerals are leached. The results also suggest that short-term increases observed are a reflection of some form of information that's stored in elixirs after only 1-3 days.

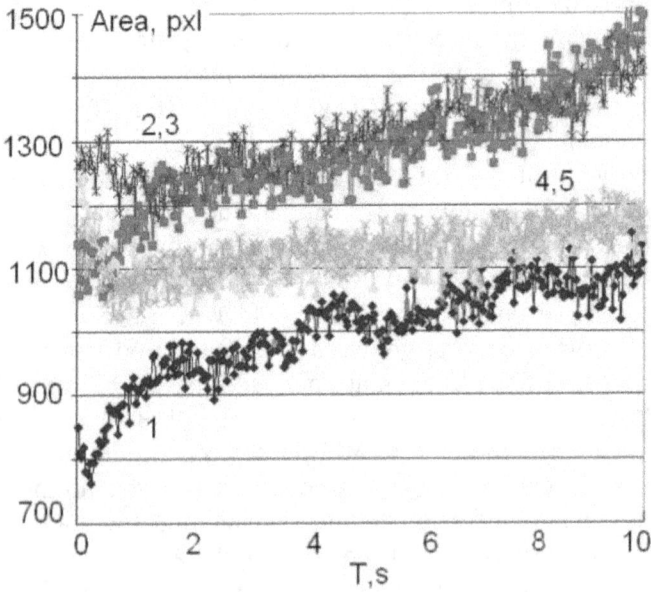

Fig. 24. *The time dependence of the EPI glow area of the water drops before and after tourmaline crystal was immersed into water.*
*1 – initial signal; 2,3 – signal after 5 min and 30 min; 4,5 –signal 1h and 2h after.*

According to homeopathic research, another experimental approach to further validate the observation of information storage, is the use of succussion. Therefore, a final series of experiments were conducted using citrine elixirs with the gemstones removed, which were measured before and after succussion as a function of storage time. These results demonstrated that in all cases, the GDV area measures show a statistically significant increase after succussion. Although the chemical content is constant in these samples, the process of succussion will introduce more nitrogen and oxygen.

There are many speculations about the influence of pyramidal structures on different processes, and, in particular, on the human condition. We developed a paper model of a pyramid, measuring 30 cm high (~ 1 ft) under which we placed a glass with 100 ml of tap water for a night. After repeating this experiment several times, we found that, in some cases, the EPI image of water being measured had changed its appearance by the morning. Figure 25 demonstrates the typical images of initial water and the water's appearance after a night under the pyramid.

We need to direct the reader's attention to the fact that these results were not reproducible. We obtained these results just several times in many experiments and found no correlations with environmental conditions or with moon phases.

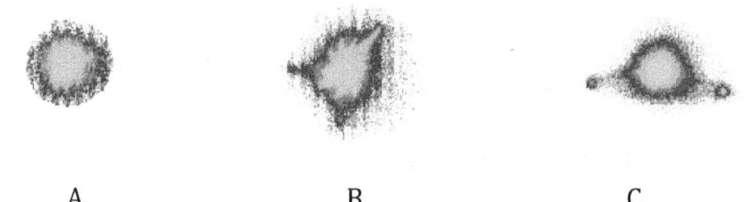

A                              B                              C

*Fig. 25. EPI images of water: A – initial tap water; B, C – the same water after a night under the pyramid.*

In another series of experiments, two closed glass vials of distilled water were kept. One was placed under the pyramid model of non-transparent glass while, in another room; another was kept in a closed, but ventilated cupboard. The Pyramid had no bottom, so air was allowed to freely circulate. After a month, the water from both vials was measured and demonstrated a statistically significant difference. Later, we were able to travel in Mexico, where we conducted numerous experiments with water kept at the top of pyramids for several hours or overnight. In many cases, we noted significant differences of EPI parameters from this water as compared with control samples.

Nowadays, the EPI method is being used for testing water and different liquids having similar chemical composition for their activity.

# "Dry Water" or "Fried Ice"

In the previous chapters, the reader could get acquainted with experimental facts proving, that besides three well-known phases (aggregation states) of water—liquid, solid, and vapor—there exists the fourth phase of water, a "gel-like," quazi-liquid crystalline, low entropy aqueous substance. The properties of this phase, formed by hydrophilic surfaces immersed in liquid water, were deeply studied by Professor Gerald Pollack and his associates who named this water Exclusion Zone water or EZ-water.

The existence of such a phase follows from theoretical considerations developed by Giuliano Preparata and Emilio Del Giudice. They applied the principles of quantum electrodynamics to condensed matter and, in particular, to water. This theory also predicts that the "gel-like," quazi-liquid crystalline, low entropy water phase can exist as interfacial water at the hydrophilic surface/liquid water boundary. "Blobs" made of water, defined by the authors as Coherent Domains (CDs), should originate and coexist in liquid water with much more chaotic bulk water.

Indeed, as A.I. Konovalov and his team demonstrated in aqueous systems—prepared by serial dilutions of solutions of many different substances to the degree when practically no molecules of the original solute can be found—there are present "nano-associates," aqueous blobs with diameters up to 500600 nm. They are rather stable and do not disappear from aqueous systems for weeks and months.

As predicted by the Preparata-Del Giudice Theory and shown by several other scientists using a wide spectrum of methods, the origination in aqueous systems of stable associates contain—not tens and hundreds of water molecules, as is usual for clusters—but millions and billions of them. American physicist of Chinese origin, Shui-Yin Lo, was probably the first to report, in the mid-1990s, data about the origination and existence of stable supra-molecular water structures in aqueous systems[48].

---

[48] Lo S-Y. Anomalous state of ice. Modern Physics Letter B, (1996). 10 , 909-919.Lo S-Y, Geng X., Gann D. Evidence for the existence of stable-water-clusters at room temperature and normal pressure" Phys. Lett. A, (2009), 373- 387.

He discovered that stable rigid "particles" appeared in the water, after serial dilutions of solutions of simple salts (like NaCl), acids, or bases of concentrations below 10-5 - 10-6 M after vigorous shaking (succussion). These particles can be seen in a transmission electron microscope with an absorbance spectrum in UV that's different from pure water. He named these structures "IE structures" to indicate that they are ice-like structures (though they can exist at room temperature). According to Lo's hypothesis, they appeared under the action of "internal" electricity.

However, Lo's data did not attract much attention until 2009, when he and his coauthors published, in *"Physics Letters,"* a paper reporting that these stable structures can be isolated from bulk water. Unlike usual water, they do not evaporate for a very long time and their properties may be studied—not in a solution, but as individual matter.

Lo and his coauthors prepared a highly diluted solution of NaCl (with all possible precautions made to exclude the presence of any extraneous particles) and dried it on a very pure glass. After the drying out process of "usual" water was completed, a spot remained on the glass that could be examined under the optical microscope (Fig. 26). Because of its very low concentration in water ($10^{-7}$ M), this spot could not originate from NaCl.

Besides, the authors distilled highly diluted solutions and dried droplets of these distillates on glass slides, in which even negligible quantities of NaCl present in original water should dramatically diminish. Still, spots of a non-evaporating substance was left on the glass after evaporation, indicating that big structures appearing during serial dilution of NaCl solutions were resistant of NaCl solutions, even to water boiling during distillation.

In this condensate, even traces of NaCl should not remain, and any associates of water present in NaCl from general considerations would have to fall apart when boiled. However, they are still present in the condensate, since after it dries out, a non-drying stain of water still remains (i.e. sustainable associates of water formed in the initial solution either did not collapse when water evaporated or had re-formed by a condensation of the vapors).

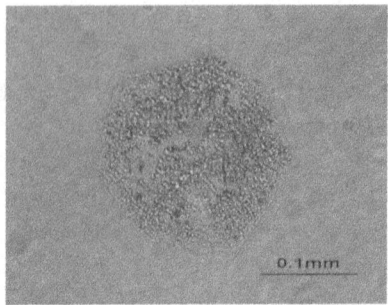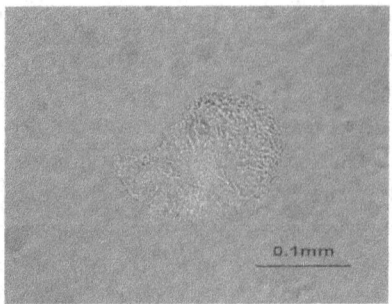

*Fig. 26. The photo on the left shows the stain left on the glass after drying a drop of a solution of 10 M g. (1a). The photo on the right, (1b) shows the spot remaining on the glass after drying a drop of the distillate solvents RA 10-7 M NaCl. (i.e. water, representing the condensation of the vapors obtained by boiling NaCl solution).*

To identify the chemical properties of the residue, an infrared spectrum of non-evaporating spots was obtained and compared with IR-spectrum of pure water. As it can be seen in Figure 27, two spectra are very similar, especially at the position of the major peaks, though the spectrum of the matter of the non-evaporating spots is much better resolved and has additional features, which indicates a much higher orderliness of this matter as compared to usual liquid water.

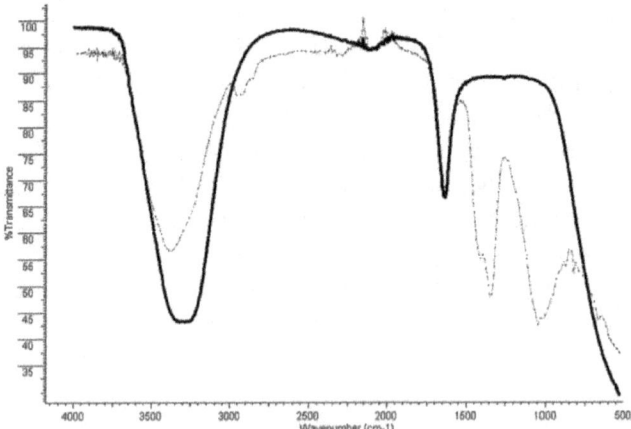

*Fig. 27. Infrared spectra for pure water (solid line) and of a spot obtained after drying of a sample of 10–7 M solution (broken line).*

In any case, the major stuff constituting non-evaporative spots is water. Yet, this water is similar to solid ("dry") compounds that do not evaporate and thus, it may represent a paradoxical species of "dry" water. The absorption peaks for pure water are 3283.5 cm−1 and 1634.5 cm−1. The absorption peaks for the non-drying sample are 3371.4 cm−1, 1639.5 cm−1, 1342 cm−1, 822.5 cm−1.

This stuff has some other peculiar properties. Using the usual optical microscopy or atomic force microscopy, one could see regular patterns on spots—a linear arrangement intersecting at the angel 102°, well-known angle between two hydrogen atoms and oxygen atom in a water molecule. The authors suggest it as a manifestation of the scaling law—fractal geometry propagation from molecular to micron scale (Figure 28).

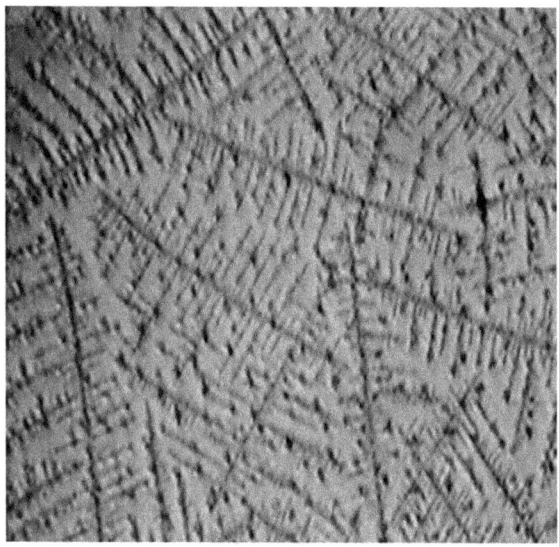

*Fig. 28. The picture of the residue left on the glass plate after drying out of a drop of 10–7 M NaCl is from an optical microscope. Another curious property of supra-molecular structures was disclosed using an Electric Force Microscope (a variant of an Atomic Force Microscope), which allowed a map of the distribution of local electric charges on the surface of macroscopic clusters to be drawn. The map revealed that the surface of large water clusters are strongly negatively charged (electrical potential at some points of the surface may reach -200 mV) and the potential differences between some points (being at a distance only fractions of microns) may reach more than100 mV.*

The emergence and existence of very stable giant water clusters in bulk water was also convincingly demonstrated by a team of Italian chemists and physicist, leaded by Prof. Vittorio Elia from the University of Naples. For many years, V. Elia and his colleagues studied the physical, chemical properties of aqueous systems of very high (homeopathic) dilutions of biologically active compounds. They proved that there is a difference in many physical and chemical properties from the ultra-pure water used in their preparation—in particular, in pH and electrical conductivity[49].

The procedure of preparing highly diluted solutions included multiple shaking of each subsequent dilution. So they questioned whether multiple mechanical treatments ("iterations") of very pure water could, by itself, change its properties. So they studied the effect on water of different iterative procedures: multiple repetitive water filtering through sintered glass filters; immersing natural hydrophilic polymers, such as cellulose derivatives (or synthetic ones, such as Nafion), in water; stirring the liquid for several minutes, then extracting a polymer sheet and drying it at room conditions. The two steps were iterated several tens of times.

It turned out that all iterative procedures produced large variations of the specific electrical conductivity and pH of water. For example, the pH changes after iterative treatment of water with Nafion. Cellulose could reach 2-3 units and specific electrical conductivity could increase by 3 orders of magnitude over the value characteristic for original water.

After iteration procedures, many other interesting transformations of water were observed. In pure liquid water, certain stable "clouds" originated whose dimensions could reach many microns. These "clouds" could be identified using polystyrene carboxylated micro-spheres (0,2 mkm diameter) that were labeled with a fluorescent dye and absorbed to aggregates having obviously different properties from bulk water (Figure 29).

Elia and his co-authors made another important discovery regarding the properties of iteratively treated water, in which

---

[49] Elia V. Germano R. Napoli E. (2015). Permanent Dissipative Structures in Water: The Matrix of Life? Experimental Evidences and their Quantum Origin. Curr Top Med Chem 15: 559-571.

stable aqueous "clouds" appeared. Using Circular Dichroism Spectroscopy, which reveals chirality, they discovered that macromolecule or supra-molecular aggregates present in this water are chiral. Chiral structures are objects, non-super-posable on their mirror image—like right and left hands. Most biological organic molecules, such as amino acids and sugars are chiral. This means that they are represented by two mirror isomers (enantiomers), one of which is able to rotate incident light to the right and another, to the left.

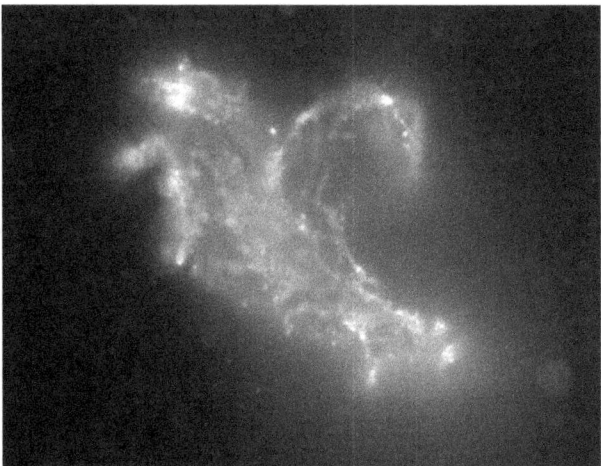

*Fig. 29. This image shows the structure appearing in water that's iteratively treated with Nafion. It was observed due to binding it to polystyrene micro-spheres marked with a fluorescent dye. In the control, Milli-Q water polystyrene micro-spheres were randomly scattered and did not aggregate with each other.*

In living organisms, only one type of optical isomers is present—in particular, left-rotatory amino acids (L amino acids) and dextrorotatory (D) sugars. Polymers made of chiral monomers, for example. Proteins are also chiral; and typical naturally occurring proteins, made of L amino acids, are known as left-handed proteins. However, water molecules, H-O-H, are achiral. That's why the discovery that supra-molecular water aggregates have optical activity like biopolymers made of chiral monomers is such a big a surprise.

The fact that chiral structures made of water may be present in aqueous systems may have far-reaching consequences. Optical purity of major bioorganic molecules—such as amino acids, sugars, their numerous derivatives, biopolymers made of them—is one of the major and unique properties of the living state of matter. After the death of the organism racemization gradually occurs. Indeed, the degree of racemization of amino acids (the ratio of D to L amino acids) enables one to estimate how long ago the specimen died.

A subject of much debate is the origin of homo-chirality in biology and the mechanism of its maintenance in living organisms. Most scientists believe that "choice" of chirality was purely random, though this "belief" is not even a scientific hypothesis because there is no way to prove or reject it. On the other hand, if water that (for some reason, not understood until now) becomes chiral as a result of such a natural process as filtration through sand, than it becomes an optically active medium. One could expect that, in the course of chemical reactions proceeding in such water, some enantiomers will take advantage over the other and the organic stuff in this system will become more and more optically pure. The advantage of such a hypothesis (over the belief in random choice) is that it can be experimentally confirmed or rejected.

Chiral supra-molecular structures present in water turned out to be rather stable. Heating chiral water at 90°C for 1 hour reduced the degree of its chirality by less then one half. Then, only after 4 hours of heating, the water lost its chirality and had the same CD spectral properties as MilliQ water. It turns out these structures are also resistant to drastic changes in pH (from pH 3 to pH 13). Such stability allows these superstructures to be isolated from bulk water. After freeze-drying different types of water—water iteratively treated with Nafion, water iteratively filtered through the glass filter, water prepared as a very high dilution of a biologically active substance using succussion (all of them having chiral properties), a solid deposit was always left in a flask (Figure 30). Its quantity was on the order of 25 mg from 250 ml of water. As was expected, after freeze-drying the control (the original pure water), no residue was left. Surely, the authors made a thorough chemical analysis of iteratively treated waters only to find practically nothing, except H2O. To be precise, water iteratively treated with Nafion (a perfluorinated polymer containing sulphonic acid groups), contained about $10^{-6}$ M of $HSO_4^-$ and $F^-$ ions. But this quantity is negligible in comparison to the quantity of the solid residue left after freeze-drying perturbed water.

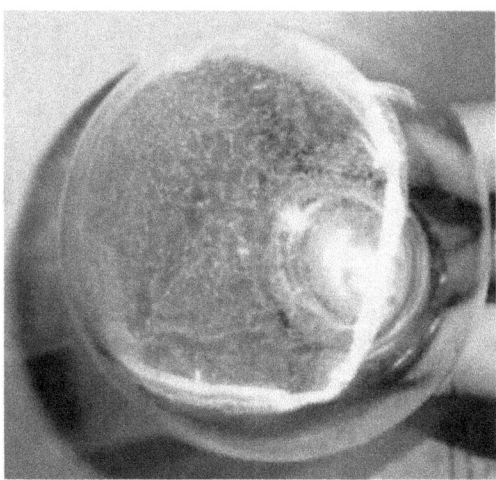

*Fig. 30. This image shows the solid residue left in a flask after freeze-drying 250 ml of water iteratively treated with Nafion.*

Using mass-spectrometry, the chemical composition of a solid residue after freeze-drying of iteratively treated water was studied. Water dominated the molecular species in the residue and its OH residue (from water decomposition) was determined. What is somewhat surprising is that this residue contained quantities of $N_2$, $O_2$, and $CO_2$. These are the very gases that should have been completely evacuated from the residue during the course of freeze-drying. From this, one may suggest, that solid residues having chiral properties may not be just some special aqueous structure, but a complex compound containing tightly bound ordinary atmospheric gases—nitrogen, oxygen, and carbon dioxide.

Long ago, complex stable compounds of water and very hydrophobic gases, such as methane and propane, were discovered and are known as "clathrates." They look like snow that does not melt at positive temperatures. However, to the best of our knowledge, clathrates of water and ordinary atmospheric gases were not described yet. It is very important that supra-molecular aqueous complexes originating in water in the course of it iterative treatment contain such clathrates, because clathrates have very interesting properties, including high biological activity.

In particular, more than a half century before, the Nobelist, Linus Pauling, suggested a molecular theory of general anesthesia. According to Pauling, anesthesia is attributed to the brain's formation of minute hydrate crystals of the clathrate type. However, he did not mean clathrates containing normal atmospheric gases[50].

Solid residue has another amazing property. As already mentioned above, chiral supra-molecular complexes in water are rather resistant to heating. But a striking result was obtained when the thermal stability or evaporation of dry aqueous residue as a function of temperature was determined using the method of thermo-gravimetry. (Figure 31).

It turns out, only when the temperature exceeds 600°C, doe all matter of the solid sample completely decompose or evaporate. Taking into consideration, that the major constituent of this solid residue is water, it may be viewed—at temperatures above 100°C—as paradoxical: "fried ice." Interesting also is the shape of the thermo-gravimetric graph above 100°C. It indicates that the ordering of material staff (HO + ?) in thermo-stable solid residue is not uniform.

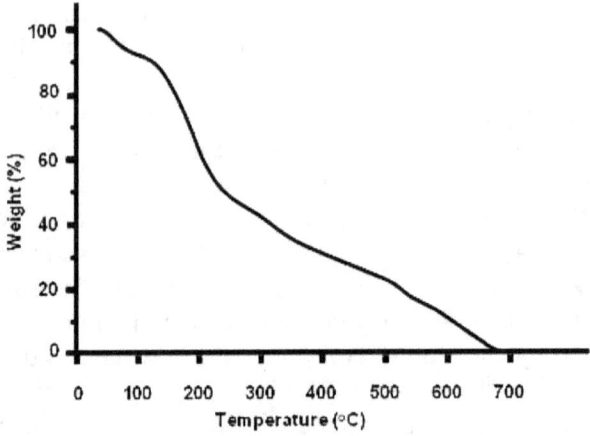

*Fig. 31. This thermo-graphimetric graph shows the solid residue left over after lyophilizing water iteratively treated with Nafion. The percentage of weight loss of the solid sample is plotted as a function of temperature.*

[50] Pauling, Linus (1961) A Molecular Theory of General Anesthesia. Science, 134, 3471. pp. 15-21

We devoted much attention to these newly discovered forms of aqueous systems, which can be quizzically labeled as "dry water" or "fried ice" because this stuff can originate in rather mild conditions and can probably exist in natural conditions. Besides, this stuff emerges when homeopathic remedies are prepared and is most likely important for their biological activity. Extrapolating the properties of Konovalov's "nano-associates" to this thermostable chiral "clouds" or "snow," one can predict that the origin of this substance and its properties should be dependent on the environmental electromagnetic fields. So this may represent the presence of a link between water and electromagnetic fields. An acknowledgement of such a link's existence provides a deeper understanding of Nature, and in the first place, of Living Nature.

What's also fortunate, is that there already exists the physical, theoretical basis for understanding all these newly acknowledged properties of water that seem so mystical within the frame of classical physics. Giuliano Preparata and Emilio Del Giudice first applied the basis of Quantum Electrodynamics theory to water. Currently, the Israeli quantum physicist, Tamar Yinnon, who has already predicted the existence of "dry water" and "fried ice," is actively developing this theory[51].

---

[51] Yinnon TA., Elia V., Napoli E., Germano R., Liu Z-Q (2016) Water ordering induced by interfaces: an experimental and theoretical study. Water 7:96-128

# Essay about Homeopathy

*The highest ideal of cure is the speedy, gentle, and enduring restoration of health by the most trustworthy and least harmful way.*
*The symptomatic palliative mode of treatment directed towards a single symptom is to be rejected.*

Samuel Hahnemann, the Founder of
homeopathy (1755-1843)

When discussing the Benveniste drama earlier, we have already mentioned homeopathy and the violent disputes that have revolved around this treatment for over two hundred years. In recent decades, this problem has gained a wider meaning and is now designated as "highly diluted aqueous solutions."

According to the academician, Alexander Konovalov: "Today, thousands (!) of examples are known, obtained in different laboratories of the world and relating to all levels of biological organization     (bio-macro-molecules–cells–organs–organisms–populations), where it is shown that aqueous solutions of biologically active substances (BAS) are able to display bio-effects (some of the latching response of a biological system) at different concentrations, including "regular" $10^{-3}$-$10^{-7}$ M (unobjectionable) and also in the area of high dilutions $10^{-12}$-$10^{-20}$ M (which raise a lot of questions and doubts). Between the two, there is a 'dead zone.' The solutions were prepared by consecutive dilutions. Hence, the term 'high dilutions' is used and given values of concentrations are estimated."

There are several directions in homeopathy. One direction is the use of substances in ultrahigh dilutions for treatment, claiming the effect of their impact to be much stronger. In this case, the doctor operates similar to a medieval alchemist: by repeated dilution of the initial substance, subsequent shaking (potentiation), maintained without the influence of bright light. Obviously, modern homeopathic manufacturing uses machines.

Despite the worldwide acceptance and spread of homeopathic treatment methods, Orthodox science and medicine remain skeptical and often hostile to homeopathy. For the rational person, it is difficult to understand how the effect could increase with dilution. It's contrary to our logic. For a man who has received a scientific education, even in the most prestigious universities, it is even stranger—in classical physics and chemistry. There simply are no mechanisms that adequately explain this phenomenon.

This leads to the fact, that in relation to such phenomena, the world's divided into two unequal camps. One camp believes in homeopathy and successfully use it in their everyday life and practice, while the majority either are not familiar with the subject or categorically deny it. In science, there's even greater separation. While their efforts are denigrated as quackery and pseudoscience, there's a relatively small group of scientists and doctors who attempt to understand this phenomenon. Their detractors don't even want to acknowledge the published papers of the major scientific journals, which present new results of research regarding the clinical effects and physicochemical properties of high dilutions.

For many, veterinary medicine's successful application of homeopathy demonstrates a significant proof of homeopathy's efficacy. In the treatment of animals, many vets (some of them we are familiar with) use both conventional methods (pharmacology, surgery), in combination with homeopathy—which, in many cases, may be more effective than traditional therapies alone. Moreover, the therapeutic effect of homeopathic medicines on dogs, cats and horses cannot be explained away by autosuggestion, or by the placebo effect.

Scientific opponents of homeopathy object to the lack of experimental evidence pertaining to the differences of high dilutions from pure water. Most importantly, they cite the complete lack of proper explanations about the physical mechanisms that can produce such effects. Quasi-geometric models of aqueous solutions are not suitable for this purpose but nothing more sensible has appeared in the last few decades. What we are addressing in this chapter is the creation of a new scientific paradigm, in terms of experimental study and conceptual understanding.

We do not aim to present a review of thousands of papers published in this area—they are easy to find on the Internet— especially for Russian-speaking readers, because a lot of interesting work had been done in Russia. (www.biophys.ru).

# Water in the XXI Century

So — finally, we have an adequate theory describing the behavior of water!

This theory is confirmed by experiments. More and more, such experiments are being conducted. Amazing new properties of water are discovered and the efficacy of ultra-high dilutions is proven. The possibility of information transfer through water can be demonstrated. This allows homeopaths to breathe more freely and spread their arms wide. Also, numerous inventors turn up to propose — through the Internet — different types of structured water with unusual properties and devices for its production.

The process did not start today. In the last few decades, the water market started to boom. All over the world, inventors began to market a large variety of devices and technologies for water processing. Typically, the inventors expounded upon the great effects of using their wonderful devices, but few make compelling cases. Although, as we have seen in previous chapters, the evidence we've received using feasible technologies and strict research protocols, include the influence of structured water to people.

People often turn to us with similar requests, and we apply various methods of research to the challenges posed. In many cases, the result is zero. Yet, sometimes we get very interesting results. When it comes to developing devices for the manufacture of good water, we probably have a much clearer understanding about the key principles it should be based upon.

• First, the initial water must be clean—not necessarily from a well half a kilometer away, but purified from harmful impurities.

• Second, it must contain a set of useful minerals and a certain level of bicarbonates. As shown in the works of V. L. Voeikov, the bicarbonates are the basis for the creation of good water.

• Water needs to be able to move—to boil, to flow, to explode, to spray, to twist into whirlpools in full accordance with the conclusions of the theory and the homeopathic practice.

• Magnetic or electric fields can influence water.

• Light has a huge impact. As demonstrated, in the works of G.

Pollack, light can create good water—as well as kill it. Currently, studying a variety of spectral sources of light will open up a huge field for further experimentation.

• All experiments are related to the influence of the time of the year and the geophysical environment. Water responds to the moon, the sun, the changing of the seasons. Thus, the same procedures can lead to different results during the spring and in autumn.

• Water becomes structured in the presence of crystals and some minerals (for example shungite).

• Water is affected by the geometry of the vessel in which it is located, not to mention the material from which the vessel is made. A good wine will never be served in a plastic bottle. Silver vessels in the Middle Ages prevented many people from epidemics.

• Information transfer makes it possible to create the structure of coherent domains in water. But this process, as we have seen in the experiments of Montagnier and Konovalov, has many peculiarities. (But if you just
set two bottles with different waters next to each other, not the fact that something would happen. The result requires a thorough review).

• The conditions of water's storage and transportation are of paramount importance. It's one thing to use some device at home to consume freshly prepared water, and quite another thing, if this water is produced elsewhere and transported over long distances.

• Then, at last, there's the influence of a person. We have no doubt that this is a significant factor. In winemaking, it is well known that some individuals produce very fine wines, while others, only table wines. As we know from many experiments, people exist who can affect water—some, merchandizing it as healing. Yet, even a simple man can charm the water simply with his positive mood or through prayer.

There are so many possibilities. For the specific device, all that's required is testing and experimentation. In our practice, we are repeatedly confronted both with water and with devices—both, which produce some very interesting results. We do not plan to cite any brands or websites. We do not make comparisons among the different manufacturers themselves. Nor do we have the desire to endorse anyone in particular. We are merely reporting a few results.

For every device, we estimate water before and after the device by using the EPI-method. We measure pH and conductivity, as well as look into how these parameters change over time; this allows access to the stability of the created structures of coherent domains (if there are any).

The most important point is to assess the biological effects of structured water. For this, it is necessary to conduct experiments on microorganisms, plants and laboratory animals. The technique for such experiments is well established. Only then can experiments on people begin.

The easiest experiment is to compare the human energy field condition before and after drinking a glass of structured water. Often, the results are very revealing.

At the 2014 Congress, "Science, Information, Spirit" in St. Petersburg, the French researcher, Guy Londechamp, presented results of his 5 years of research —to influence people to drink a glass of water prepared in his device. The device was a geometric construction—a copper tube 16 mm in diameter, having left turn portions in a rising part while descending part was divided into 2 symmetrical divisions (with torsion bar synchronized steps aimed in opposite directions) which connected in the resulting matrix.

After testing more than 100 people, Londechamp found that for 80% of participants, the energy of their field increased, and for many, their chakras became balanced (figure 32). The latter is an indicator of their psycho-emotional balance. (It should be noted that the energy growth may be observed only in people who have relatively low energy in the initial state. When all is well – there is little to improve). Below are a few examples.

Naturally, the question arises: how long does this effect persist? The measurements show that after a single glass of water, improvements last from half an hour to an hour. But this does not mean that it is necessary to drink water each hour on the day. It makes sense to drink a few liters during the day—although this depends on many other factors: age, weight, climate, exercise and so on.

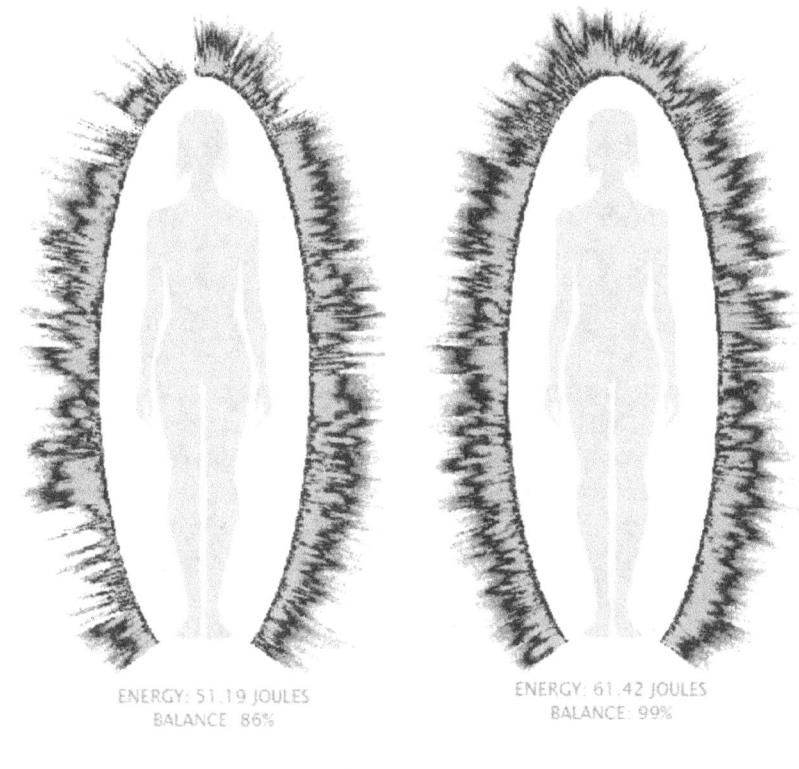

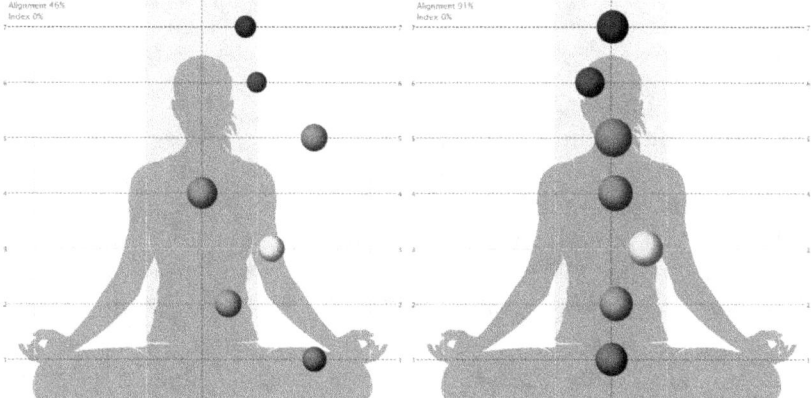

*Fig. 32, A change in the energy field and balance of Chakras after drinking one glass of structured water.*

A group of specialists from the St. Petersburg Research Institute of Physical Culture and Sport (including medics and psychologists) conducted a serious investigation about the long-term effect of consuming structured water. The study was conducted on athletes living and training at the St Petersburg school of the Olympic Reserve. These schools were organized during the Soviet era; the young guys are studying and practicing on a full academic scholarship. Many of them later become Olympic and world Champions. The advantages of such cohorts are that they are all about the same age, healthy and undergo regular medical examination, living and training in the same conditions, eating the same food. Therefore it is easy to organize randomized groups in which all participants have roughly the same physiological state.

Forty boys and girls participated in the study. Using complex techniques, their physiological parameters were measured in their initial state. Then they were subjected to physical loading: 10 minutes of active work on the bike while doctors measured recovery parameters after exercise. Then the whole group was randomly divided into experimental and control groups. In the experimental group, participants for a month drank only the structured water from a special filter, while the control group only drank regular bottled water. The subjects knew about the purpose of the experiment but were not informed of the type of water they would drink.

By the way, one of the advantages of working with athletes is that in the training process, they drink a lot of water. A month later, the measurement process with loading was repeated. The results were clear: there was a statistically significant difference in physiological parameters of control and experimental groups. The authors of the study came to the following conclusions:

1. The results of the experiment show that after a month of drinking structured water, athletes experienced statistically significant changes in the parameters of their cardiovascular system. Heart rate parameters and diastolic blood pressure decreased after exercise, while recovery time following exercise also decreased by 18%. This data indicates improvements in physical performance, optimization of the circulatory system, and enhanced exercise tolerance. Such trends were not observed in the control group.

2. Based on a variety of pulsometry data, members of the

experimental group had a tendency towards the optimization of the vegetative balance (increased parasympathetic effects on the Heart Rate Variability parameters and decreased sympathetic ones). This attests to the improvement of the capabilities of the body.

3. In response to exercise, athletes in the experimental group demonstrated an increase in the values of their mental strength factor, reflecting the level of competitive readiness.

4. Data obtained using the GDV method suggested that the values of energy parameters for the athletes in the experimental group remained stable, whereas the control group exhibited a decline in the values of these parameters. At the same time, in the experimental group, there were significant increases in the energy potential values pertaining to specific organs and organ systems.

In conclusion, the results of the experiment demonstrated that athletes drinking water (that had passed through a special filter), experienced growth in aerobic capacity, physical performance, energy potential, along with their bodies' adaptive reserves.

The result is convincing. It clearly shows that if you constantly drink good water, the body could much more easily cope with the strains of everyday life. Good water, in particular, is one of the prerequisites for longevity. Naturally, this fully applies to animals and plants.

Before a cat, two bowls of water were placed—one with tap water and the other with specially treated water. The response was clear: the cat chose the structured water. This also applies to plants.

A large experiment was conducted in India under the leadership of scientists of Tamil Nadu Agricultural University (Bangalore). Farmers were provided with special devices developed by an Australian-American company, in which water flows through a tube, which contains natural crystals and glass spheres, creating toroidal vortex water flows. This movement mimics the flow of water in mountain streams. It's the opinion of the inventors, that this type of water flow gives water special dynamic and structural properties. Water, after the device has been tested with the GDV device, showed a statistically significant difference between the original and structured water .

Local Indian farmers were asked to water some of their fields with structured water, and some with normal. The result exceeded all expectations.

## Conventional Water          Structured Water

| PLANTS | Conventional water | Structured water |
|---|---|---|
| **The acreage** | **0.375 acres** | **0.375 acres** |
| **Wheat** | | |
| # of grains on the stalk: | 180 good; 56 bad | 310 good; 50 bad |
| Harvest: | 355 kg | 640 kg |
| **Zucchini** | | |
| Plant Height: | 25 inches | 35 inches |
| Amount of branches: | 1-2 | 3-4 |
| Harvest: | 1100 kg | 2250 kg |
| **Tomatoes** | | |
| Harvest: | 1326 kg | 2042 kg |
| **Cotton** | | |
| Plant Height: | 23,7 cm | 35,5 cm |
| Amount of leaves: | 1326 kg | 2042 kg |
| Length of the leaf: | 4.49 cm | 9.02 cm |
| **Beans** | | |
| Harvest per bush: | 0.702 kg | 1.458 kg |
| **Green Peppers** | | |
| Harvest: | 38.5 kg | 68.7 kg |

Equally impressive results were obtained on a poultry farm.

| CHICKENS | Conventional Water | Structured Water |
|---|---|---|
| # of Chickens: | 2430 | 2970 |
| Total Weight: | 3864 kg | 5528 kg |
| Average Weight: | 1.59 kg | 2.20 kg |

The benefit of structured water not only sowed in an increase of the chicken's overall weight, but the following moments were also noted:
- Not a single case of chicks death of was reported.
- Antibiotics were not used.
- Chickens looked healthier and were more active.
- All chickens were around the same weight. All these results were obtained only through the use of non-expensive systems of water activation!

A lot of our "new" is well-forgotten "old!" We have already mentioned the research of Viktor Schauberger, an unrecognized genius, as well as that of Rudolf Steiner, on the influence of shape on the properties of water. But at that time, these works were not supported and were forgotten in time. Only now, a hundred years later, they are finally seeing proper development. The English engineer, John Wilkes (1930 – 2011) in 2008, created a Fund for Water Foundation, which aims to help water to support life.

On the basis of the principles of fractal geometry, they design and build, in stone and ceramics, biodynamic geometric shapes which can be symmetrical, to organize the flow of water. They are not designed for private gardens, public spaces or industrial production. In all case, they are based on principles developed by Wilkes and his followers.

Dynamic water is used in the manufacture of products and processes with improved quality without incurring additional costs. An important application is the use of biodynamic designs for the cleanup and revitalization of water. The figure shows the result of using such structures for cleaning lakes in New Zealand. As Schauberger said, water has the property of self-healing—you just have to let her do it.

*Examples of the biodynamic structures of the Foundation.*

Photographs of a lake in New Zealand showing before and after photos of biodynamic treatment.

Since 1979, the Fund has performed over 5,000 projects in 79 countries. It is interesting to note that after such treatment, around the reservoirs the number of mosquitoes became sharply reduced. The Fund holds a large volume of research on plants and biological crops. The results are always revealing. The figure 33 shows an example of the stems of coriander grown in the same conditions using the conventional and biodynamic water. In biodynamic conditions, the plant acquired the ability to fractal self-organization—it changed the shape of the leaves and increased its root system. The plants also became more viable and more effectively coped with pests and diseases.

One more interesting application of water in the XXI th century, is the novel idea to transfer the properties of drugs on the water via the Internet. Many good scientists and researchers developed this idea and, in the beginning of the XXI st century, a nonprofit organization was founded In Moscow under the name "The DST

Fund." The goal of this organization has been to promote innovative developments of Soviet and Russian scientists in the field of non-conventional informatics.

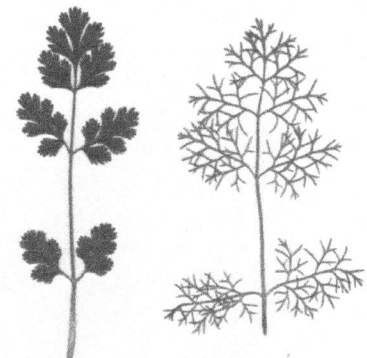

*Fig. 33. An example of the stems of coriander, grown in the same conditions, using the conventional and biodynamic water.*

The General Director of the Foundation, Eugene Germanov, is a businessman and investor, with 20 years of experience. The Fund owns a number of patents for various innovative technologies. In 2009, the Fund started the project "Emission"—the transmission of drug properties through the Internet, while retaining all its properties. Since then, hundreds of tests and experiments have been conducted. Time after time, evidence has been collected that the properties of a substance can, indeed, be transmitted through the communication line thousands of kilometers.

The DST Fund cooperates with 23 doctors from Russia, the Ukraine, Lithuania and the USA and scientists from many Institutes of the Russian Academy of Sciences and Moscow State University. Many experiments and pilot clinical tests demonstrate the efficiency of the technology. Patients were able to reduce the dosage of medications. In some cases, they were even able to totally substitute their medications with the informational copies. At this time, downloading informational copies of medical drugs is free of charge (www.dst-fund.com), although The Fund plans to move this process on to a more commercial footing.

# Water as a Detector of Emotions

*I don't want to be at the mercy of
my emotions.
I want to use them, to enjoy them,
and to dominate them.*
Oscar Wilde (1854-1900)
*The Picture of Dorian Gray*

The EPI/GDV-bio-electrography method allows registering the stimulated glow of water that depends on its structural state. This fact has been meticulously studied and proven. Numerous studies have shown that the structural state of water changes under the influence of various factors. A human's directed attention is among them.

In our experiments, we registered the GDV-glow of water that was remotely influenced by the operator. We have been interested in this subject for a long time, since the end of the 1990s when the methods of studying the parameters of the GDV-glow of water were developed. In the middle of 1990s, together with the famous sensitive Alan Chumak, we conducted the first experiments of this kind. Through concentrating his attention, Chumak changed the state of water, (detected from the changes of the signal's glow). At that time, we developed a good method for measuring the signal of glow: from a meniscus of liquid suspended above the electrode of the device. This method remains the most sensitive method for studying the gas-discharge glow of various liquids: water, blood, oil and other solutions of various substances.

The Bible tells us that Jesus turned water into wine. Was it Cabernet or Merlot? That seems unlikely. Water remains water—but maybe it can obtain special properties? What is wine's the most valued property? Why has wine always been so appealing to mankind?

The main property of wine is to make people merry. After having a drink, people forget about their everyday troubles, the world seems exciting and significant, and everyone becomes happier and jolly. People want to sing and dance. Wine is a very special laughing liquid.

Therefore, Jesus did not turn water into something else, but he bestowed very special properties into it. After drinking that water, people became merry and happy. They sang and wanted to sing praises to Jesus. Speaking in present-day terms, Jesus structured the water by directional increase of its biological activity. Numerous experiments—regarding water being mentally influenced—show that, on the whole, this is a distinct possibility.

Sitting only two meters from the experimental installation, Alan Chumak concentrated upon a column of liquid. He became so absorbed in this process that he did not notice anything around him. At that moment in time, no sounds could disrupt his mind's focus.

Special measurements indicate the moment an individual enters a specific state of altered consciousness. The conditions and study of this subject constitute a separate branch of research. Achieving altered states of consciousness is common for a shaman, a healer, a surgeon (performing an operation), and even in the case of a distinguished actor performing on stage. In this trance-like condition, a person can and often does influence the processes around him.

In the majority of such experiments, statistically significant changes of the parameters of the water glow were detected. The glow image became more active, and the number of ramified streamers grew. But the greatest changes observed were for the dynamic glow curves that were changing over a prolonged time period—up to several days. These results were not observed each and every time—they depended greatly on Chumak's ability to consistently achieve an altered state of consciousness. This process may not always be governed by our will alone.

Further experiments—studying the influences upon water's glow parameters—were conducted with the cooperation and participation of many "sensitives" from different countries. In many cases, the experiments demonstrated the effects of their mental influence were statistically significant.

Interestingly, after the first experiments where the operator mastered the specifics of the task, the influence could be performed, from then on, at virtually any distance.

Another series of experiments studied a group's collective influence upon water. The experiments were conducted during seminars in groups of 20-30 students that had a positive attitude towards the professor and the subject of discussion. After explaining the purpose of the experiments, and registering the initial images of the water glow, individuals were asked to use their inner concentration to send feelings of love and goodness to the water. In many cases, the experiments succeeded in changing the parameters of the glow of water.

We conducted a large number of similar experiments. In the majority of cases, we received positive results. These convincing results show that the parameters of the water glow undergo statistically valid changes, whether under the influence of the directed attention of a single human or a coordinated group of individuals. The distance between the operator and the device is insignificant, be it two meters or two thousand kilometers.

In that case, the experiment is conducted in the following way: several sealed bottles containing the same water are placed on a table in a laboratory. Colored felt-tipped pens mark the bottles. The operator attempts to influence only one bottle of his/her choice, at a random time within 2-3 hours. The operator writes down his/her choice, places the note into an envelope and seals it. When the specified period is over, the glow signal is measured for all bottles. All measurements are repeated five times, whereupon statistical indices are calculated in order to evaluate the difference in glow of the different samples (see example at figure 34). Clearly, in the absence of a significant influence, all samples must yield statistically identical results. Different results may indicate that there was an influence. After completing the measurements and preparing the results, the final result is prepared as a protocol and is signed by all participants of the experiment, whereupon the operator's envelope is opened and the results are compared. Thus, the experiment becomes a double-blind study: none of the participants know what the operator is going to do.

We conducted similar experiments with operators from different countries. Among them were Alan Chumak and Aleskey Nikitin from Russia; Christos Drossinakis and Victor Filippe from Germany; and Krishna Madappa from the USA. In all experiments, the operator was first informed about the conditions of the trial. Then they attempted the experiment at close range. Only after that, would they work in the double-blind mode from a long distance.

Sessions between Germany and St. Petersburg (Russia); Moscow and St. Petersburg; and Japan and St. Petersburg were performed. The efficiency of the influence at a relatively close range (from an adjoining room) for five of the above-mentioned operators was

80%. In three experiments of the long-range action study, 10 test trials were performed with the overall efficiency of 70% (7 successful tests out of 10; in 3 cases, the signal from all bottles was the same); in 5 cases, the color chosen by the operator matched the color of the bottle where changes were detected; in another test, changes were detected in two bottles, and yet, in another case, the intentions of the operator differed from the actual results.

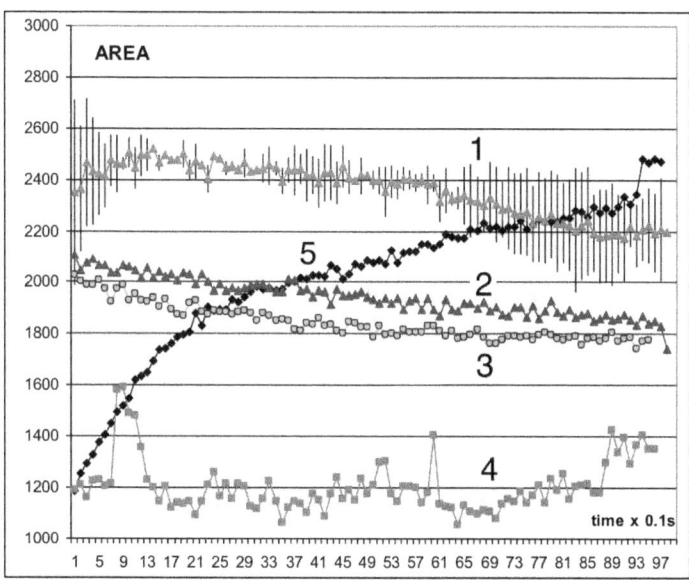

*Fig. 34. The time dynamics of the EPI Glow for Water Drops from Five Different Bottles. Every one is averaged on five subsequent measurements. Vertical bars at the upper curve demonstrate the typical level of variation. As we see, curves 1, 2, 3 and 4 have quite stable dynamics, while curve 4 is variable. From this, we make conclusions that in bottles 1, 2, 3 and 4, water was intact while water in bottle 4 was influenced by the distant intention.*

The influence of healer, Alexey Nikitin, from the distance 25 meters resulted in the transformation of relatively stable water dynamics into chaotic oscillations (Figure 35). Under the influence of Valerii Sochevanov, a very similar effect was initiated.

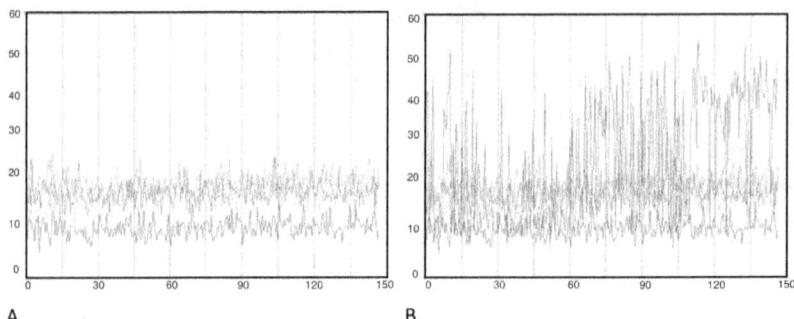

A.                                    B.

Fig. 35. *The Time Dynamics of the area of EPI Glow for water samples before and in the Process of Distance Influence.*

From 2007 to 2010, a series of experiments with remote intention influence were conducted together with Lynne McTaggart. Lynne organized large groups of people through the Internet (http://theintentionexperiment.com). People could see the photo of the experimental setup and began their meditation at a specific, pre-arranged agreed upon times. People were not given any special instructions as to any method of meditation. Data was recorded continuously, in automatic mode, about one hour before the intention time and one hour after. The control experimental setup was set up in an adjusted room. Recorded responses of both water installation and EPI sensors were statistically significant ($p < 0.05$). In all cases of collective directed intention, big groups of people from all over the world recorded statistically significant responses of water and sensors. Although control samples demonstrated some variation of parameters, it was not significant in most cases.

Experiments of this kind are performed in the double blind trial mode and the data is statistically processed, thus excluding any possibility of random changes. The control measurements (performed in the absence of any deliberate influence) in all cases showed that the water in the bottles produced an identical signal.

The selection of water is crucial to the success of the experiment. Over time, different samples of water have different stability of the glow signal. In the majority of cases, during the registration of the glow, the signal grows for 3-5 minutes and then stabilizes at a certain level.

Distilled water has the smallest stabilization time, and its signal is extremely stable, albeit low—but distilled water barely reacts to any mental influence. That is why it is rarely used in experiments like this.

The best reaction to mental influence is manifested by natural water, but one must first check that the glow signal is stable and reproducible. Stability can be achieved by obtaining water from a natural source and keep in the open air for 24 hours. Usually the experiments use natural water that's been bottled and sealed and only opened upon the commencement of the experiment.

The above-mentioned sensitive operators each used their own internal methods for creating a measurable influence. All of them reported they established mental contact with water as if it were a living thing, and then transferred emotion to this thing. Therefore, we can safely say that water acted as a detector of emotion.

### So what are the main conclusions from these experiments?

Have you ever noticed that a dinner cooked at home by a loving spouse is more delicious that the most exquisite food from an expensive restaurant? And don't forget your mother's trademark dish that she makes just for you. Could it be that sincere and kind feelings a person feels during the cooking process influences the structure of water and changes it—as happened in our experiments? By the same token, food prepared with ill feelings may equally produce extremely negative results.

In India, the Hindu culinary tradition pays special attention to this particular issue. Ayurvedic texts state that prior to lighting the cooking fire, one must pray to reach a positive frame of mind and only then, proceed with the cooking. If you cannot discard the negative emotions and hard feelings, it is better to skip cooking this time. We should always remember: all our emotions and feelings influence the water around us, all the time.

# Water as a Material Matrix
# for the Consciousness Field of Humankind

*I sent my Soul through the Invisible,*
*Some letter of that After-life to spell:*
*And by and by my Soul return'd to me,*
*And answer'd "I Myself am Heav'n and Hell.*
Omar Khayyam (1048-1131)

**The XXIst Century is the time of creating a new science of consciousness.**

Consciousness is the ideal category—it's the basic imperative—along with Matter and Information. In modern science, we do not have the notion of consciousness, either in medicine or in biology. We've only undertaken the first baby steps in understanding this concept. Obviously, even transitioning to micro- or macro-worlds involves changes in ideas about consciousness because it is difficult to talk about expected behaviors. From a bird's eye view, people's behavior on the streets of a city, observed in time-lapse, is curiously reminiscent of an anthill or a beehive.

Our first step in science is to provide a definition. Often, people argue for hours about certain topics, defending different positions. People only keep in mind their own understanding of the topic. But in reality, they are all referring to different things. So let us begin with defining what we understand by the term "Consciousness."

History attempts to build a rigorous, but not intuitive, definition of consciousness. Rooted in the distant past, all these attempts are inevitably based on an *a priori* division of the whole of nature into two parts: either conscious or unconscious. The definition formally draws the border between them to be exactly where expected. Depending on this, the authors of the concept make statements about the adequacy of the specific wording. Definitions based on this division, are subjective and, biased, most often in favor of Homo sapiens. With this approach, avoiding logical tautology is not possible in principle.

In our studies we use the following definition:

**Consciousness is a property (ability) of natural objects to form abstract representations of the outside material world,**

**suitable for use in purposeful activity through sensory perception.**

In more simple form:

**Consciousness is the ability of a subject to react to environmental information and change its behavior in accordance with this information.**

From this point of view, every living organism possesses consciousness—all animals, large and small, above ground and underneath it, in the air and in water; plants; microorganisms; plus every cell of our bodies. At the same time, inanimate subjects do not possess consciousness: a stone heated by the noonday sun cannot move or change its color to avoid overheating.

But what about robots—according to this definition, wouldn't the best of them possess some type of consciousness?

From our standpoint, robots can only replicate certain elements of the consciousness of their creators. In the case of rapid changes, living beings are able to modify their circumstances, even costing the death of part of the population. Of course, to some extent, out of millions of years during the kingdom of dinosaurs, only crocodiles were able to survive. Robots may behave in accordance with a program, designed by engineers, and their ability to modify is quite limited—at least, within the limits of known technologies.

All discussions above are related to **the first, basic levels of consciousness.**

**The next level of consciousness is the ability of a subject to predict future events, remember past events, plan and modify its behavior to meet future situations in accordance with their experiences.**

This level is characteristic of higher animals and humans. All owners of cats and dogs know that their pets may be very smart when it comes to not only their favorite food, but often to their relationships with their masters. A book by Rupert Sheldrake, *"Dogs That Know Their Owners are Coming Home"* (1999) presents interesting examples. Seven experiments demonstrated that dogs were able to accurately predict their masters' return and were waiting near the door. We hear many stories about the conscious behavior of dolphins and elephants, not to mention monkeys and crows. All mentally healthy people have this level of consciousness, albeit to different extents. Children naturally possess this level at the age of two to three years of age.

**The next level of consciousness is the ability to generate new ideas, exchange these ideas with others, transmit them to the next generations and manifest them in objective reality.**

Only human beings possess this ability. Ants have built their communal and living areas in the same way for a millennia—as do bees, birds and many other animals. We cannot explain how they achieved this ability in the first place, or who taught them, but they never altered the structure of their constructions. Very rarely do they introduce new elements acquired through painstaking random trials.

In the case of humans, creativity is the basis of the development of civilization. Artists, poets, architects, scientists, and many other creative people have advanced our civilization. The whole of today's modern world is created only through the constant generation of new ideas. Some people generate new ideas, while others turn them into reality. Many smart people reach this level of consciousness in their work, which allows them to make everyday decisions, based on knowledge, experience and intuition. All successful business people, politicians, teachers, engineers, builders and many other professionals have to generate and rely on new ideas in their everyday practice. The innovative process is so common people often do not pay much attention to it. But it's only in this way, that humankind can create and develop new advances in our civilization.

**A high level of consciousness is the ability to generate ideas—not based on the existing level of social development and knowledge—which transforms society, moving it to the next level of civilization. Presumably, connecting to the Universal Information Field and receiving direct information from the Higher Planes of Consciousness accomplishes this.**

We can name particular people known to us in history, who were able to attain this level. First of all, stand our Great Spiritual Teachers: Moses, Zarathustra, Buddha, Jesus and Mohammed. They were able to provide people with new understandings of their lives. They formulated spiritual laws, which has governed millions of people for thousands of years. In different words, in different languages, they were all speaking about the same things—about human life, the human soul and the role of humans in the Universe.

Let us understand, that all religions basically provide the same message. All have similar spiritual content, which points to the same deep meaning. Great Teachers were able to present people complicated spiritual truths in simple, understandable words. Within a short time, their teachings have been accepted and embraced by millions of people, because the teachings allowed every individual to reach new levels in their spiritual development. Moreover, all these teachings have existed for thousands of years. These ideas do not die; they transform in accordance with the development of society.

There is no need to mix Spiritual Teachings with human institutions. Religious organizations are social structures, organized by particular people in specific socio-historical periods. They were initially designed to help people with their everyday needs, to help them survive harsh conditions, explain spiritual ideas in understandable words. However, all too soon, these structures were transformed into social institutions, which have been used in the Power Games of Rulers. Sadly, we know of many crimes perpetrated under the name of God—millions of people have been tortured and killed by religious fanatics, in religious wars, and this process still continues today.

All we discussed above relates only to individual consciousness. Humans also possess a **Collective Consciousness**. We seem to be the only creatures on Earth, who have both an individual and a collective consciousness. For the creation and development of Civilization, this is our most powerful tool. Like neutrons in the brain, we link our individual minds, and thus increase our group's intellectual power by a thousand times. The formation of civilization became possible only when people started living in large groups—when they formed the first cities, they generating extra products, which allowed more and more people to be freed from the everyday struggle for survival. By combining their brains and their abilities, people created Civilization.

Of course, our definitions are only some of the possible ways to discuss consciousness. You may find many other different approaches in modern science. Our aim is to give definitions, which we may useful in our research, and allow us to develop experiments, which may help us to understand some enigmas of human existence.

It is noteworthy that the rigid structure of a collective, being surprisingly stable for ants and having ensured their survival for hundreds of millions of years with all the changes in the environment, turn out to be totally unsustainable models for human society. Even being forcibly established and guarded, such structures spontaneously disintegrate in an historically short time. History shows that Empires exist for about 1000 years, before they disintegrate, to be replaced by a new Empire. We observe this pattern throughout the whole of human history, regardless of nation or continent.

Some social formations exist for much shorter periods—mere seconds from an historical perspective. The socialist system existed for 70 years, and quietly broke up without any pressure from outside. It is obvious, that with the suppression of the Individual Consciousness in the name of the Collective in human society, the fundamental laws of life of the noosphere, the laws of the stability of a complexly organized system are violated. The Communist model for the structure of a society is perfect for protozoa and insects; however, all attempts to introduce it to people—to beings with a more advanced sense of organization—have been shown to be unviable.

We have not mentioned such notions—obligatory in modern psychology—as "unconsciousness," "subconsciousness," and "superconsciousness." There are thousands of books discussing these topics. We presume that readers understand their meaning. The reason is this: these are descriptions of brain functioning. They describe the interrelations between different parts of the brain, particularly the left and right hemispheres, in the processing of received information. It is a well-known fact that we take in much more information, than we are aware.

The brain processes all the information it receives but it presents to our attention only a small fraction of it. This process depends on each person, the situation at the time, and the importance of the information. Actively intuitive people may process much more information from the environment as compared with those who don't normally exercise their intuitive skills. In our definitions, this is related to activity associated with different levels of consciousness. I would say, this describes the inner mechanisms of informational processing. In our definitions, we are more focused on the outcome of this process.

Figure 1 presents our main ideas on the Levels of Consciousness. Group consciousness is typical—both for human and for animals—but also with ants, bees, termites, fish, etc. At this level, individual elements cannot exist without the group and group behaves like a separate organism consisting from many separated particles, bound together by collective bonds. Tribal systems are organized on this principle as well.

Collective consciousness (collective unconsciousness after Jung) is the driving force of human civilization. Here, every individual has his/her own highly developed consciousness. But it's only by combining their consciousness by the **Collective Consciousness Field,** that we create conditions for our development. By this Informational Field, we understand all possible means of informational exchange: verbal, printed, transmitted by electromagnetic fields, via Internet, or transferred by quantum effects.

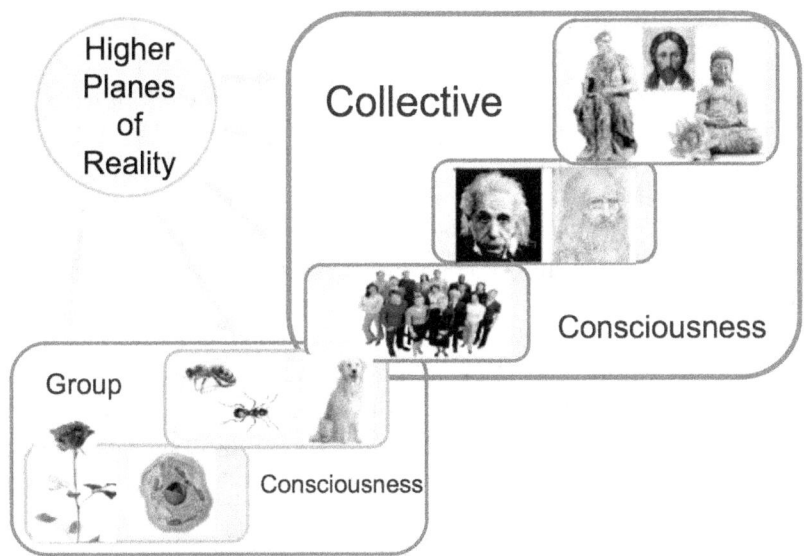

There's a lot of evidence that people can voluntary influence their own physiological processes. They can exert influence, at a distance, upon physiological and psychological state of other individuals and may also remotely impact material processes. The evidence is mostly anecdotal or data that's has not been well studied. Another aspect of these phenomena is the influence of the collective consciousness on the material world as compared with the influence of individual consciousness. Jung first proposed the idea of the collective consciousness. Since its inception, the idea has been strongly explored in psychology and the social sciences. There are quite a few experimental proofs of the remote effects of the collective consciousness upon material processes.

At the same time, the Collective Consciousness Field is responsible for such effects as: Inspiration, Precognition, Reincarnation memories, Healing, Telepathy and much more. You may find a lot of scientific, experimental data on these effects published in peer-review journals.

### But, what is common to all conscious beings? It is organized water!

As we demonstrated in the previous chapters, water has awareness; water has memory; and water has foresight. These are the very same properties that describe our notions of consciousness. In humans, consciousness is most evidently associated with the human brain. Recent developments in neuroscience demonstrate that changes in the states of consciousness are attributed to changes within the activity of different parts of the brain. The main conclusion is that consciousness (and memory) of an individual is correlated to brain neuronal activity. This is the ruling paradigm in many human related scientific fields.

At the same time, some of the most widely used methods for monitoring brain activity—in particular, functional MRI and diffusion MRI—primarily register the state and dynamic activity of that water, which constitutes the overwhelming majority of brain matter (which, presumably, depends on nervous activity and reciprocal influence).

By far, the brain is the most morphologically complex form of living matter. So, what unique properties does this organ possess?

The brain is one of the most "wet" organs of organisms. Water, in the brain, is most assuredly, sophisticatedly and dynamically structured.

Water constitutes the majority of brain matter. As in any other tissue, it is represented by both extra-cellular and intracellular fractions. But inside the brain, specific states are achieved, on the one hand, by an exclusively complex architecture of nervous and auxiliary cells. The predominant fiber of their branching structure ensures, at any moment, a specific structural organization of significant fractions of water. On the other hand, oxygen-dependent metabolic processes proceed in CNS much more intensely than in any other tissues and also provide for the permanent, highly excited states of brain as living matter.

Continuous and highly organized (coherent) changes of structural—energetic states of aqueous component of brain matter may provide brain with the property of being both the receptor and emitter of informational signals (in particular, but not exclusively, of an electromagnetic nature). Properties of the brain have been visualized, thanks to the most sensitive technique — functional Magnetic Resonance Imaging (fMRI). This method directly evaluates the state of water in cells and tissues. Dr. Raymond Damadian invented MRI in 1971, and based it on Gilbert Ling's idea of water structuring in living cells.

To introduce consciousness into a scientific framework, we need only develop a theory of its operation, which would be able to explain different levels of consciousness and predict some new phenomena—such as distant healing, mediums and out-of-body experiences. At the moment, the most promising work seems to be with theories based on quantum electrodynamics.

We need to accept consciousness as a system effect. This system is not depend upon any particular part of the body, even one as powerful as the brain. Its dependent upon the entire system working as a whole. We may attribute some level of consciousness to every cell of the body, or to an organ in particular. But to achieve a high level of consciousness, we need the coordinated activity of all the cells along with all of the organs working together.

Quantum electrodynamics operates with the notion there's a system consisting of many elements. That is why it may be applied to the construction of the theory of consciousness. You can find many papers, published on this topic. However, without being a mathematician, it is absolutely impossible to understand or appreciate their importance and meaning. In this area, interested readers may first refer to the works of Mari Jibu, Kunio Yasue, Emilio Del Giudice, and Giuseppe Vitiello among the list of other prominent theoreticians.

The application of quantum field theory describes why and how classical behavior emerges at the level of brain activity. The relevant brains states themselves are viewed as classical states. Similar to a classical thermo-dynamical description arising from quantum statistical mechanics, the idea is to identify different regimes of stable behavior (phases, attractors) and transitions between them. This way, quantum field theory provides formal elements from which a standard classical description of brain activity can be inferred.

And, as has been shown in previous chapters, G. Prepagata and Emilio Del Guidice developed one of the most convincing theories of water, which they based on quantum electrodynamics.

Another amazing fact (using data from the Herschel Space Observatory) is that astronomers have detected, for the first time, cold water vapor enveloping a dusty disk around a young star. These findings suggest that this disk, which is poised to develop into a solar system, contains great quantities of water. It further suggests that water-covered planets, like Earth, may be common in the universe.

Herschel is a European Space Agency mission with important contributions from NASA. Scientists previously found warm water vapor in planet-forming disks close to a central star. Evidence for vast quantities of water extending out into the colder, far reaches, where comets take shape, have not been seen until now. The more water that's available in disks for icy comets to form, the greater the chances that large amounts of water will eventually reach new planets through comet impacts. "Our observations of this cold vapor indicate enough water exists in the disk to fill thousands of Earth oceans," said astronomer Michiel  Hogerheijde of Leiden Observatory in The Netherlands. Hogerheijde is the lead author of a paper describing these findings in the Oct. 21, 2011 issue of the Journal, *Science*.

There are several worlds thought to possess liquid water beneath their surfaces, and many more worlds that possess water in the form of ice or vapor. Within our solar system, water is found in primitive bodies, like comets, asteroids, and on dwarf planets like Ceres. The atmospheres and interiors of our four giant planets—Jupiter, Saturn, Uranus and Neptune—are thought to contain enormous quantities of the wet stuff, plus their moons and rings have substantial water ice. Scientists, using NASA's Hubble Space Telescope, recently provided strongly convincing evidence that the five ice moons of Jupiter (Ganymede, Europa and Callisto) and Saturn (Enceladus and Titan) are water worlds. All show powerful evidence of oceans lying beneath their surfaces.

It's easy to forget that the story of Earth's water—ranging from gentle rains to raging rivers—is intimately connected to the greater story of our solar system and beyond. But our water came from somewhere—and every world in our solar system, likewise, got its water from the same shared source. So it's worth considering that the next glass of water you drink could easily have been part of a comet, or an ocean moon, or a long-vanished sea on the surface of Mars.

And note, the night sky could be full of exo-planets formed through similar processes as our home world, where gentle waves wash against the shores of alien seas. [http://www.nasa.gov/]

At the same time, ideas that our Consciousness can expand outside the Earth's boundaries into the Cosmos was first proposed by Russian Cosmism—an existential philosophical orientation that views the survival of mankind, and of the individual, as an integral part of humanity's "common task." The migration of humans into space is regarded as inevitable, since this move is essential for humanity's long-term survival. An increase in the lifespan of humans points to achieving another essential task.

Among the major representatives of Russian Cosmism was Nikolai Fyodorov (1828–1903), an advocate of radical life extension by means of: scientific methods, human immortality and the resurrection of dead people. Konstantin Tsiolkovsky (1857–1935) was among the pioneers of theoretical space exploration and cosmonautics. In 1903, he published the book, *"The Exploration of Cosmic Space by Means of Reactive Devices (Rockets),"* the first serious scientific work on space travel. Tsiolkovsky believed that colonizing space would lead to the perfection of the human race, with immortality and a carefree existence. Other cosmists include Vladimir Vernadsky (1863–1945), who developed the notion of noosphere, and Alexander Chizhevsky (1897–1964), the pioneer of "helio-biology" (the study of the sun's effect on biology).

If we accept, to some extent, an idea close to Cosmism, what would be the material carrier of Consciousness in Space?

The answer is easy. Water in the Universe may be the material carrier of the Universal Consciousness!

Many experiments show that human consciousness can directly affect water. Under the influence of human consciousness, water can change its properties and these changes may be quantitatively evaluated and transformed to other media—both by direct changes may be quantitatively evaluated and transformed to other media, both by direct contact and digitally.

We suppose that water, being the most abundant substance in

the environment, may serve both as the channel for information transmission and reception and as the major component and processor of information storage. This hypothesis, when proven, will challenge the ruling paradigm in neuroscience, medicine, health care, water science and philosophy. It will help to generate a significant shift into a new world scientific paradigm—a new scientific revolution! As we see, there are many common properties human consciousness and water share. We can list some:

| | |
|---|---|
| Consciousness is Omnipresent | Water is Omnipresent |
| Consciousness receives and stores information | Water receives and stores information |
| Consciousness is described by Quantum Field Theory | Water is described by Quantum Field Theory |
| There are different States of Consciousness | There are many different states of Water |

All of the above-mentioned allows us to propose this hypothesis: that **water serves as the material carrier of the individual and the collective Informational Field of Consciousness.**

This hypothesis should be tested with further research.

# Part II

# WATER and our HEALTH

# Introduction

*"We drink 80% of our diseases"*
Louis Pasteur

**"WATER IS LIFE"** is as true for non-humans as it is for humans. No wonder the earth's natural waters teem with various living organisms—many of which are dangerous for humans.

At the same time, the inexorable statistics show that 80% of all diseases in the world are caused by an unsatisfactory quality of drinking water and violations of sanitary and hygienic standards of the water supply.

The so-called "water diseases" can be divided into four groups:
• Diseases caused by infected water (typhus, cholera, dysentery, poliomyelitis, gastroenteritis and hepatitis)
• Diseases of skin and mucous membrane caused by using contaminated water for drinking and washing (from common pimples to trachoma and leprosy)
• Diseases caused by water-dwelling mollusks (schistosomiasis and dracunculiasis)
• Diseases caused by vector insects that live and breed in the water (malaria, yellow fever etc.)

Modern water supply systems often malfunction. When this happens, it directly causes outbreaks of water-related diseases. American researchers believe that the main cause of many epidemics is insufficient purification and disinfection of water. More than 2 million people die every year from malaria alone. Water-dwelling mosquitoes carry the disease. Large-scale studies conducted in several Russian cities (http://www.iceberg-aqua.ru) reveal statistically valid connections between the number of people asking for medical aid in relation to gastrointestinal diseases (in particular, gastric ulcer and duodenal ulcer), and the potable water quality indices. The most significant statistical correlations of these diseases were found to be with the concentrations of iron, heavy metals, and the color and turbidity indices in tap water.

Specialists have detected the influence of drinking water with increased hardness, increased contents of sulfates, chlorides, nitrates and cyanobacteriae on the development of urolithiasis and cholelithiasis, functional disorders of stomach and allergic diseases.

"The results of the scientific studies indicate that due to intensive pollution of surface water bodies and disturbances of the ecological balance, the water-dwelling microorganisms secrete resistant toxic substances that affect the nervous, immune and digestive systems of human organisms and cause mutagenic aftereffects."

Note that this was neither stated by a journalist looking for a fantastic headline, nor by a fantasy horror fiction writer. The Specialists of the State Committee officially provided this data for the Sanitary and Epidemic Monitoring of Russia.

In addition, the problem of the formation of toxic organ chlorine compounds (including dioxins) in drinking water during its chlorine disinfection is becoming more and more urgent. But the list of problems facing ordinary consumers of potable tap water in their own kitchens is still not complete. The fact is well known that we receive up to 25% of our required daily amount of mineral substances with water. And these mineral substances have a significantly higher physiological value than those supplied with food.

Any surplus (as well as regular deficit) of a chemical element can easily transform water from friend to foe. Here is an example: a surplus of sodium chloride (the principal factor in water mineralization) above 1g/l (0.12 oz/gal) results in increased reactivity of blood vessels, deposition of salts on the vessel walls (the main reason for strokes and heart attacks!) and disturbances within the water-salt balance of the human organism.

One can hardly argue with the importance of iron for correct functioning of the human organism. This macro-element is an essential component of hemoglobin and myoglobin and it is a part of cells and enzymes. But even ancient people knew that a cure and a poison often differ only in the dosage. Medical researchers observed that long-term intake of water with increased content of iron increases the risk of strokes, affects our skeletal system and has a negative effect on reproduction. Moreover, dry and itching skin is also an indication of an iron surplus in the body.

However, the consequences of excessive intake of other macro- and micro- elements are no less dangerous: increased concentration of copper affects mucous membranes, kidneys and liver; surplus of nickel affects the skin; surplus of zinc results in kidney disease. An accumulation of chromium, lead and cadmium promotes the development of oncological diseases and disorders of the nervous system. Intake of water with high contents of boron and bromine results in diseases of the digestive system.

In the last few years, scientists have been actively discussing the role of aluminum in the development of Alzheimer's disease. Concentrations of aluminum in water above 0.5 mg/l or 0.5 ppm are proven to significantly increase the mortality rate from this disease. Such concentrations were demonstrated to have an inhibitory action on the central nervous system and the immune system of children. A high concentration of fluorine in water (limit value is 1.5 mg/l or 1.5 ppm) tarnishes the teeth with stains (fluorosis), while insufficient concentration is a high risk factor for cavities in the teeth.

Also, the hardness of water, which was not regarded seriously by the medics until now, is drawing a lot of attention. Extremely soft water increases the mortality due to cardiovascular diseases. The risk of chronic nephritis and hepatitis, increased mortality, toxicoses of pregnancy and congenital malformations come from drinking water polluted with nitrogen-bearing and organochlorine compounds.

Of course, this list of dangers to our health from the intake of low-quality water is far from complete.

# Water and Hygiene

*In rivers, the water that you touch is the*
*last of what has passed and the first of that*
*which comes; so its time present.*
                                    Leonardo da Vinci
                              from his Notebooks, c. 1500

Any process can run to its extreme. Thus is the case with water. The industrial era brought water to every house, and made bathtubs and showers publicly available. A lot has been written about the beneficial effects of dousing oneself with water, especially cold water, and no one doubts the hygienic importance of washing. We consider bathing, bathtubs and showers to be an integral part of our lives. But this was not always the case.

Bathtubs, swimming pools and bathhouses were an essential attribute of civilization in Ancient Greece, Rome and the oriental countries. We are all familiar with the Roman Thermas and Turkish baths—an inevitable practice for countries with hot climate, it would seem. However, the custom was entirely opposite in the Mongolian steppes, where summers are far from being cold. The Yassa of Genghis Khan—the principal law under the Mongol Empire for hundreds of years—specifically stated that bathing in summer time was punishable by death for warriors. We can only speculate about the reasons behind this prohibition, but the most logical explanation is that a warrior jumping into a cold river on a hot afternoon would be left without his clothes and weapons, therefore providing an excellent opportunity for attacking enemies. We should remember that the lives of Mongols at that time were devoted to constant warfare, with both external and internal enemies. We can only imagine how much those Mongolian warriors reeked during the summer!

However, Medieval Europe was not much better. Here is how Georges Vigarello described the situation in his book *"Le propre et le sale: L'hygiène du corps depuis le Moyen Age (Concepts of Cleanliness: Changing Attitudes in France since the Middle Ages)"*:

*From the XV century, every outbreak of plague resulted in physicians condemning bathhouses and pools where naked bodies came into contact with each other. People afflicted by contagious* diseases mix there, which cannot help but alarm us. The danger of catching a disease there is great: "I beg you to stay away from steam rooms and bathhouses or face a certain death." However, temporary closure of bathhouses and steam rooms that always took place during plague outbreaks was consistent with the isolation logic. Starting from the XVI century, the bathhouses began to be regularly and officially closed. For instance, by decree of the Provost of Paris (that was renewed several times between the epidemics of 1510 and 1561), prohibited visiting steam rooms and prohibited their owners to heat them up until Christmas, under threat of a punishment. Such decrees were enacted in a growing number of towns. The prohibition was spreading: in Rouen it was enacted in 1510, in Besançon – in 1540, and in Dijon it had been in force since the end of the XV century.

The first purposeful actions against plague—especially from the beginning of the XVI century—create a horrifying image: the body consists of permeable membranes. Its surfaces do not prevent the entry of water and air, and its borders become even vaguer when faced with an ailment whose material foundation is not visible to the eye. We cannot discard the possibility that the pores are weak per se, and their weakness is only partly caused by the heat. The pores should be constantly guarded from hostile influences. That is why the cut and quality of clothes is very important during epidemics: the clothing should be smooth, densely woven and should fit the body very tightly.

Humors escape through the pores, and consequently, so do the forces. These holes are open for two-way motion, and the internal substances seem to be extremely eager to escape the body... That is why washing makes us feeble and stimulates dementia. It exhausts the forces and the virtues. In other words, the hazard is not only in the infection. Such notions become popular enough to breach the boundaries of a medical discourse and become part of the mentality. Washing without obligatory precautions is unthinkable: relaxation, bed rest and protective clothes are indispensable. Washing is a certain cause for concern. It becomes more and more complicated and rare, since the number of precautions increases, but the complete safety still remains unachievable. Vulnerability of pores is alarming in several directions: they intersect and continue each other. First of all, hot water affects a passive body, permeates it and leaves it open and vulnerable.

Due to similar concerns, the practice of washing babies is discontinued fairly early. Otherwise the organism, wet as it is, will remain slack and limp. Washing can hinder the gradual drying of the flesh that is the growth. The material will remain too soft. From the moment when the baby begins to look neat and rosy, it is too dangerous to resume the bathing. For instance, the dauphin (the future Louis XIII) never had his legs washed above his feet until he turned six. After a short bathing when the baby is born, the next submersion in water does not take place until the age of seven.

When books about health published in the XVI century mention certain bodily odors, the need to eliminate those orders is also stated. However, using ointments and perfume in such cases is considered preferable to washing. The skin should be wiped with a perfumed cloth: "When the armpits stink, one should gather a bunch of roses and rub them into the skin." Rub diligently to impart a pleasant odor, but never wash.

Wearing a perfume when going out is not just an aesthetic gesture. And strolling with a lump of ambergris in your hand is not just an attempt to follow the fashion. Here is a very typical description of Paris of Henry IV made by an Italian traveler, the description that remained accurate for decades: "A stream of fetid water runs through all city streets, absorbing sewage from each house and poisoning the air: that is why people must carry flowers – to mask the stinking." Another significant report is the difference between several hospitals in Paris observed by a Bolognese traveler Locatelli that was curiously observing France of Louis XIV: for instance, the atmosphere at Hotel Dieu had a remarkable stench, and there were four and sometimes five patients in each bed.

So movie scenes, where characters from medieval France bathe in a tub of water, are the product of the imagination of storywriters. This is why foreigners visiting Russia were so amazed to witness naked people emerging from the bathhouses and leaping into the river or even an ice hole.

# When and how should we drink water?

*Man lives by 75% on the basis of their fantasies and only 25% - based on the facts.*
*Erich Maria Remarque (1898-1970)*

The man is 75% water.
School Biology

In order to feel good and to ensure the optimum functioning of our body organism, we must observe our water consumption schedule. We react very strongly to variations of water content in our organism and can only survive without water for a couple of days. After losing about 2% of our body's weight in water, we begin to feel thirsty; after losing 6-8%, we enter the near-unconscious state; and after 10% of lost water content, we begin to hallucinate. Water content loss above 20% is fatal for humans. To illustrate this, I present a story from my climbing practice days.

*Once, our sports team of four decided to climb the high altitude peak in the Pamir Mountains. This was during the Soviet times when I was a member of the mountaineering sport team of the Soviet Union. At the time, we were participating in the national championship in mountain climbing, and our task was to go through several challenging peaks to gain fitness.*

*After the first few ascents, we decided to conquer one of the peaks by a steep wall nearly a half a mile in height. In the previous year, this ascent took the prize in the Championship. We read the description of the route: the climb of the wall was supposed to take three days, plus a day's descent to the camp. In the report, it was written in particular, that there would be plenty of ice on the wall sections (still at the height of over 17,000 feet above sea level), and that there'd be no problems finding water. Therefore, we only brought a certain amount of freeze-dried food along with gas stoves to melt ice and for cooking.*

*The first day was very active—the wall had a southeastern exposure—so that from morning until nightfall, it was hit by sunlight and climbing on dry rocks was very nice. In the evening, we reached a narrow rock ledge where we could spend the night sitting. The only problem was—we did not find ice. Apparently, that summer was* warmer than the previous one. We just wanted to drink and had no appetite. So we did not raise the primus, drank 2 tablespoons of water from a small jar and attempted to sleep.

The next day was painful. All morning, the sun was shining and even at that height, its rays heated everything all around. We experienced terrible thirst. According to an old remedy, we rolled a stone in our mouths, but that recipe did not help. Our tongues swelled. We had to periodically shift our tongues from side to side with our fingers. At the same time, we still had to climb the steep cliff. To establish the rope, we had to perform our tasks with full concentration and attention to detail. By the afternoon, we were able to come together on a small ledge. All of us looked terrible—but in fact, we still had another day of hard work ahead of us!

And then someone pulled out, from a backpack, a small bottle of Tajik balm—an herbal tincture based on alcohol. "Look, I was going to present it at the top, but maybe we can try," he said. We took a tablespoon each, and it was like a heavenly balm.

Our feelings of euphoria were amazing! It was as if a fire broke out in the mouth and filled our mouths with delicious flavors. Swallowing was not a problem—this was a balm (it responded to this word), which was completely absorbed in the mouth. We immediately felt a surge of strength. "Let us have more," someone suggested. "No, we have just a little bottle, let's save it for the evening" became our joint decision. We made it up, but we never encountered ice.

We spent the night on a little rock shelf drinking a spoonful of balm. Nobody even mentioned food. We slept fitfully, dreaming of lakes and fountains of fresh water.

In the morning, it all continued. Still, there was no ice. Our rate of movement had fallen sharply; our heads were pounding and our vision blurred. We had to be doubly careful. Suddenly, in the middle of the day, one of our companions hissed: "I hear the creek around the corner—I will go and look..."

He began to unfasten himself from the rope to leave to go somewhere. These were hallucinations—at this height, there would be no running water. Water would only be in the form of ice and snow. We tried to persuade him, but he was so eager to leave, we decided to give him a spoonful of balsam to calm him down. That's how we figured to keep him under control. I climbed a rope, fixed the belay, and shouted to him: "Get up on the rope, here you will get a *spoonful of balsam." It worked and he furiously climbed up to receive it. But the balm's effect ended all too quickly.*

*In our state of mind, none of us were clear. Should we hold on to the top? Our minds were in such a daze and each movement was difficult. Naturally, it was the dehydration with the added effects of the high altitude. But, according to our estimates, due to our loss of pace, reaching the summit was still a day's climb away.*

*The evening was falling. I climbed the next rope. Above me was a small rock balcony. I pulled myself up and climbed to the balcony. And there I found that I was lying on ice hidden from the sun by overhanging rocks. The winter ice remained intact! I acted on my first impulse—to break off a piece of ice covered with small stones, and thrust it into my mouth. It was water! To take in water was such an indescribable feeling! Cheered again to life, I immediately felt a surge of strength! An hour later found us all gathered at this site: we put up the tent, lit our kerosene stove and we stoked ice. Late into the night, we drank warm water with pleasure. We never had a meal. A day later, we were back at base camp. For the next couple of days, we only drank tea, juice and water with citric acid.*

How much water do we consume? This is not an idle question as three-quarters of our body consists of $H_2O$. Humans should take in from 2 to 5 liters (or quarts) of water a day. An adult receives about 1.2 liters of water (48% of recommended daily intake) in the form of various drinks or liquid food. The remaining water, about one liter, is obtained with food.

We present some curious facts: porridge contains up to 80% of water; bread – about 50%; meat – 58-67%; vegetables and fruits – up to 90%. This means that 50-60% of so-called "dry" food is water. Then about 3% of water is produced through various biochemical processes in the organism. So, for the final 1-2 liters, we must drink.

An important fact to consider is the specific, proven benefits of simply increasing one's daily intake of water—or replacing coffee, tea, alcohol and soft drinks with water. For example, the morning cup of coffee (so enjoyed by some people) should be preceded with a glass of water. In any case, drinking a glass of clean water in the morning before breakfast is beneficial for our health. This glass of water activates almost all systems of the human organism. And, according to certain data, drinking plenty of water also encourages activity and promotes longevity.

In London, a meta-analysis of the survey data of 3427 adults, aged 20 to 80 years, was conducted, some of who suffered from cardiovascular problems and chronic kidney disease. They were divided into three groups (depending on the amount of water they consumed per day): low consumption < 2.0 liter per day; average 2.0 - 4.3 liter per day; and more > 4.3 liter (more than 1 gallon) per day. The survey result showed that people who consume very little water had a much higher risk of developing kidney disease. Any consumption of other beverages had no effect. Consuming vast amounts of water produced no relationship to cardiovascular problems.

So we can safely state that too small of a daily consumption of clean water may lead to disease, while any negative effects of consuming 4-5 liters a day has never been observed anywhere.

There are several small villages where the inhabitants boast excellent health and longevity. Mountain villages are not the only villages that have become a symbol of an optimum living environment. There's also a village near the coast of Japan, as well as a town near the Pacific coast of the USA, not very far from San Diego. What all these places have in common is their inhabitants' style of living, which is based on the laws of healthy living, as we understand them today.

These village denizens are physically active, eat healthy food, do not overeat, consume little sugar or alcohol, and in some places, eat a lot of fish. They do not suffer from obesity and remain vigorous into old age. This way of living is popular in many places. However, only the majority of inhabitants from these particular villages live for more than 100 years. As a rule, these people live in the same way as their neighbors. So what makes those villages different from the neighboring ones? Serious scientific research did not reveal any significant factors: they live in the same way and they eat the same food. The only difference is in the water they drink.

In all those special places, for their entire lives people have been drinking water from local sources. Since their villages are small, there is no need to subject the water to treatment with chlorine and other chemical substances. The local source of water completely satisfies their needs. They drink their own water, and they drink a lot of it—up to several liters per day. The products of the modern chemical industry, like soft drinks, are not very popular there.

So if you want to be healthy—drink water, not sugary drinks.

# What should we drink?

*Water is critical for sustainable development, including environmental integrity and the alleviation of poverty and hunger, and is indispensable for human health and well-being.*

*United Nations*

Numerous articles and books clearly state that drinking clean water is good for our health. There is no doubt about that. But what degree of cleanness should the water standards meet? Should we treat it by means of double osmosis, and then add the salts, or is it enough to follow the threshold limit value regulations? We found scant little experimental data published regarding the influence of water of different qualities on the state of human health. That is why we decided to conduct our own field experiment.

The project involved students of an institute in St. Petersburg who lived in a dormitory. They were chosen due to living in approximately the same conditions. The students had similar food habits and water consumption schedule, as well as being roughly of the same age. A group of healthy subjects was selected—24 boys and 26 girls. These 50 subjects were randomly divided into two groups, 25 subjects per group. Using Electro-photonic Imaging, all participants had their basic physiological parameters measured: their blood pressure, body temperature, heart rate variability and biofield parameters. No statistically significant difference was found between the original parameter values of the two groups, meaning that their psycho-physiological parameters were basically the same.

One group (the control group) continued living in the same way as before, while the experimental group received clean water, promising to drink and use only this water for cooking, at least one liter per day. No restrictions were imposed on the diet or in the consumption of alcohol and non-alcoholic beverages. Thus, the students continued with their lives in the usual way, the only difference being the fact that the participants of the experimental group drank clean water in the morning and in the evening. Once per month, all 50 subjects had their physiological parameters measured.

After only one month, the parameters of participants in the experimental group surpassed those of the control group; by the second month, the difference became even more pronounced. After three months, the difference between the parameters of participants of the two different groups became visible to the naked eye. Only two subjects in the control group could compare with the participants of the experimental group. These two turned out to be active sportsmen who took great care of their health. After excluding them from the sample, a statistically significant difference between the two groups was discovered. According to several parameters, in the experimental group the intensity of their basal metabolic rate increased, indicating an overall activation of physiological processes of the organism.

This experiment serves as clear proof of the fact that regular consumption of clean water (in the amount of at least one liter per day) during several months has a statistically significant effect on one's health. That is only the case, of course, when not consuming soft drinks.

By the way, although soft drinks have become one of the most successful businesses in the modern world, they are far from being healthy. The times when such drinks were made from natural fruits and berries are long gone. What we drink now in our society is merely a bunch of chemicals. The pleasant smell and taste of soft drinks, as well as the occasional acid color is the result of a successful synthesis in enormous chemical factories.

Once during a TV talk show, the CEO of one of the companies that produce soft drinks claimed that all ingredients used in production were absolutely harmless, even for newborn babies. Then a representative of the environmentalists stood up and asked the CEO if the preservatives used by the company were also harmless. The CEO confirmed it without doubt. Then the environmentalist showed the CEO a little bottle and said: "If it is true, then you surely will not object to drinking a glass of this preservative in front of the audience? It is after all totally harmless, even for a newborn baby!" The CEO hesitated, started to mumble some excuses, and said that these substances were only used in very small quantities. He refused to drink from the bottle. This segment was later cut from the TV show.

The reader might well argue that people drink such beverages all their lives and nothing happens to them. Why should they be harmful?

First of all, the present-day society can look upon the sad experience of the United States, the richest country in the world. Coca-Cola is the national symbol of this country, the parents even give it to their little babies, and vending machines with soft drinks can be found even in the middle of forests. The result is evident— no other country has so many overweight people as the USA. Of course, not 100% of the population is like this, some got luckier with their physiology and genetics. Individual variability and resistance certainly exist. But even for the most resistant people, chemical substances accumulate in the liver, the bones, the cells, and over time, they produce a cumulative destructive effect on the organism.

So if you want to be healthy, drink fewer soft drinks.

**Natural juices** are a different matter. But still, we should be cautious. Too much orange juice increases the production of histamine, which can cause asthma in children and adults. The natural sugars found in juice encode the liver to operate in the fat saving mode—a good recipe for becoming overweight.

And if we recall that dissolving a concentrate in water composes all packaged juices, a question immediately pops up. What kind of water do they use? None of the juice manufacturers provide data about the water they use. Usually, they use a bi-distillate—a hygienic and safe choice. But the value of such drinks for the human organism is another questionable issue, not to mention the preservatives in juices or the so-called fruit beverages, which use the same sweetened chemicals.

**Milk** should be regarded as a food. Milk is undoubtedly a good source of calcium and proteins dissolved in water, but we must not forget that the milk of a cow was intended by Mother Nature to feed a calf that starts walking several hours after being born. Adult humans often do not even have the enzymes required for breaking down milk. In particular, this enzyme is virtually absent in people of the Mongoloid race.

Everyone knows that undiluted cow's milk should never be given to newborns or babies that do not yet walk. Bottle-fed infants should receive more water. Some autopsy cases revealed cholesterol in the coronary arteries of bottle-fed infants. The long-term effect upon our organism of the chemicals used to prevent milk's souring remains uncertain. Minimal quantities of such chemicals have no effect on our health. But who knows the collective effects of regularly consuming processed milk over several decades? Such long-term experiments cannot be conducted in all the laboratories of the world.

Dairy products treated by bacteria belong to a totally different class. Cheese, cottage cheese, kefir and their numerous derivatives have served us as an important source of food for thousands of years. Prudently consume only those products that are free from preservatives and chemicals.

**Coffee and tea** are also classed as a food. A cup of coffee contains about 80 mg of caffeine; and a cup of tea—about 50 mg. Chocolate also contains caffeine and theobromine that act in the same way as caffeine. Caffeine causes dehydration within the organism, since the amount of excreted urine exceeds the amount of water contained in the drink. We are not urging you to deny yourself the pleasure of one cup of coffee per day, but you should also keep in mind these inconvenient facts.

A glass of clean water in the morning right after waking up will help you to significantly improve the functioning of your gastrointestinal tract and many other systems of your body organism. The optimum time for the ingestion of water during the day is as follows: one glass of water 30 minutes before meals (breakfast, dinner and supper) and one glass 2.5 hours after each meal. This is the minimum amount of water required for the organism. Another glass of water is recommended after a hearty meal and before retiring to bed. Some people believe that establishing a regular, properly calculated daily intake of water could prevent the development of the diseases that plague modern society. This is an exaggeration, but it errs on the side of truth.

Remember the main principle for a healthy life. Water is one of the principal sources of health and nothing can replace clean water.

There is an old belief that a person shouldn't drink anything at least an hour before and after the physical exercise. Is this true? Can we drink while engaged in sports? This belief is far from the truth and not replenishing the body with fluid can be very dangerous to our health! Our organism loses up to 2 liters of water during one hour of exercise. This results in dehydration, disturbing the body's heat regulation, overheating, and significantly increasing the risk of complications from the cardiovascular system (especially dangerous to children and old people). That is why maintaining proper water balance during physical exercise is vital! These are scientifically valid guidelines regarding water consumption and exercise:

• 2-3 hours before the exercise you must consume 400-600 ml of liquids;

• During exercise, replenish water loss by drinking 150-350 ml of water every 15-20 minutes (depending on the intensity of the exercise);

• After the exercise: replenish 150% of weight loss caused by the exercise by drinking the same amount of water;

• Risk of dehydration increases with: high temperature of the surrounding environment, as well as low humidity, and also by exercising in clothes that prevent normal heat exchange.

# What is good potable water?

*Whiskey's for drinking, water's for fighting about.*
          Mark Twain (1835-1910)

What kind of water runs from our tap? What substances does it contain? Is it safe to drink? In different areas of the country, tap water is supplied either from underground sources like boreholes or from surface sources like rivers, lakes and water storage reservoirs.

The first and most obligatory level of water quality testing is compliance with the State Standard that specifies the standard norms for various pollutants. Theses norms define a safe level of pollutant contents. If a substance is found in the water in a concentration below its threshold limit value, this water can be used for human consumption without being a threat to health. Yet, this does not indicate that this water is biologically active or good for our health.

Surface water is usually more prone to pollution. Water reservoirs can become contaminated with sewage from factories and farms, or with acid rain, or with microscopic algae and even pathogenic microorganisms that propagate in the water. Freshwater reservoirs are among those ecosystems that are most affected by mankind's activities. This is why such water is more thoroughly cleaned. Special water treatment stations filter the water, bind its pollutants with coagulating agents and, prior to supplying the water to the water supply system, sterilize it to destroy any microorganisms lurking there.

Usually, water from artesian wells is less polluted, since polluting substances from the surface cannot reach it so easily. However, such water tends to have higher contents of dissolved salts of calcium and magnesium. In other words, its hardness is greater. And such water is sometimes cleaned in a rather superficial way, relying on its natural purity. But pollutants still can penetrate very deeply through the cracks in rocks.

Water from boreholes, springs and wells should be examined with extra attention. One can never guess whether sewage waters have contaminated any water-bearing strata in the underground rocks they pass on their way. From the many underground burials of industrial waste, as well as ordinary industrial landfills and sewages, there is also the danger of leakage. The spring water resulting from this would be a source of some most unexpected components.

One never really knows where containers dating from the cold war might be buried—deep below the ground, somewhere in the woods, near an abandoned firing ground... After several decades, these containers gradually deteriorate, causing poisonous substances to be washed away by rainwater. These processes are especially active in spring, during floods. Perhaps the poison enters our underground streamlets, flowing for dozens of miles along the water-bearing horizon, until it enters a clear spring surrounded by people filling up their empty bottles. No comment.

So, what should we do to have good water? The main idea is that those wishing to be healthy and merry should take care of those themselves. Your health is in your hands. You decide what to eat and what water you wish to drink. These are the factors that greatly influence our health. Good water must first be specially prepared. Filters and water treatment methods should be used. In order to evaluate the quality of water, have a sample analyzed in a laboratory. Such labs can be found in every city. Some analyses, providing at least an approximate evaluation of the water, can be done at home by means of standard assay kits for aquariums. In this case, you will at least have some notion of the kind of water you are dealing with.

# Water Treatment Methods

*In the arid States the only right to water,
which should be recognized, is that of use. In
irrigation this right should attach to the land
reclaimed and be inseparable there from.*

President Theodore Roosevelt
State of the Union Message, Dec. 3, 1901

Obviously, potable water must be specially prepared.

What are the methods for water treatment used presently in our world today?

Certain popular books describe various affordable methods of water processing, such as freezing, infusing water with silicon, silver etc. Of course, all these methods provide certain beneficial effects, as long as you have the time and enthusiasm to implement them. Unfortunately, a working individual usually has no time to look around, let alone pour water from one pot to another. So we are discussing practical methods which may be used in everyday life.

## Settling

Water settling, performed for at least 3 hours, lowers the concentration of free chlorine, so it is important in water preparation for watering home plants or filling wish tank. All aquarists know that fish respond well to the tap water that has been settled for at least 24 hours. But settling has virtually no effect on the amount of iron ions, heavy metal salts, and other substances in water.

## Distillation

Distilled water is not fit for regular consumption since it does not contain the microelements essential for the human organism. Moreover, it washes the mineral salts out of the system. That is why frequent consumption of distilled water disturbs our immune system, heart rate, digestion, etc... The main idea employed by many water producers is to add proper amounts of artificial salts to distilled water, thus creating "sterile and healthy potable water."

Unfortunately, this idea is not working well. The "artificial water" produced lacks some component that makes natural water good for our health. Which component is that? It's probably the structure of water, the subject of numerous scientific research projects and discussions.

Of course, in some places like Dubai, where there are no natural sources of potable water, ocean water desalination is the only way to get water for consumption. A bad decision is better than no decision.

## Filtration

Selecting a good filter is no easy task. In order to understand what type of filter to purchase (and there are many different types of filters—charcoal, membrane, bactericidal, integrated etc...), one must first evaluate the composition and special properties of the water in question.

The filter needs to be selected according to the characteristics of the water in question. This is a job for a professional. A home filter is basically a mini water-processing factory. Simple filters usually remove only mechanical impurities and extra chlorine. Complex installations are designed to produce water that meets the 150 criteria defined by the World Health Organization.

Naturally, water-processing filters quickly become clogged. This lowers the degree of purification. Then the filter soon starts releasing the accumulated impurities and thriving microflora back into the water. We cannot predict the time when the reverse contamination of water may start, and therefore only regular replacement of the filter element can guarantee high-quality purification.

New devices have been created in the last few years—not just as a guaranteed way of water purification—but also as a means for making the water beneficial for our health. The innovative working principle of such devices is structuring. General water treatment principles can be embodied in your own apartment or a country cottage. To retain our integrity, we will not mention any trademarks. This book will not turn into an advertising brochure, which is why the principles are described in a fairly general manner.

The first stage of purification—the first barrier in the way of the water flowing into an apartment—is a mechanical filter. Large impurities, like sand, litter and dirt, are invariably removed at water purifying stations. But after traveling through many kilometers of rusty pipes, water gathers various impurities yet again. Naturally, it depends on numerous conditions, but a sound idea is to check your tap water by filling a one-liter glass bottle with it and letting it settle for 2-3 days. The sediment will show the extent of the pollution in your tap water. By the way, as a rule, mechanical filters are easily cleaned by running water flow through them and do not require extra maintenance.

The next stage is a multi-step absorbing filter. Several steps are recommended, at least 2 or 3. The first step consists of various fibrous filters manufactured from state-of-the-art materials. The next step is the charcoal filter, an indispensable part of all cleaning systems. This further stage consists of the micro-porous filter and other filters recommended by the manufacturers. schungite[1], silicon and silver filters perform very well at these stages.

The important thing is to remember that all these special-purpose filters, (as well as water activation systems), must be installed after the initial purification stages that eliminate dirt, chlorine and harmful impurities. Revitalizing dirty tap water does not eliminate the principal factors that have pathological effect on our health. The following guidelines should be observed when selecting and installing a filter:

- Make sure the filter capacity is sufficiently high. Each filter has a certain processing capacity, which is usually specified by the filter manufacturer. These recommended values must be strictly observed;

- The lower the speed of water passing through the filter, the better the quality of cleaning. It is best to choose purifying systems where the water slowly passes the filters and then enters a special accumulative reservoir from which it is later taken for drinking. It's good to keep schungite, silicon or quartz in this reservoir;

---

[1] **Shungite** is a black, lustrous, non-crystalline mineraloid consisting of more than 98 weight percent of carbon. It was first described from a deposit near Shunga village, in Karelia, Russia, from where it gets its name. Shungite has been reported to contain fullerenes.

- Filters should be regularly cleaned, and filter cartridges should be replaced at regular intervals, especially in the frost-free season. The scourge of any filter is the growth of microflora that starts actively polluting the water and turns a healthy device into a potential health hazard;

- You should check the effectiveness of your home water purification system by having the samples analyzed in a laboratory. This will verify the validity of your choice.

In our homes we are using three-stage filtration with commercially available filters, then we structure potable water with CrystalBlue device (http://www.crystalblueent.com); this water we pour into silver or clay jar with shungite, amber and silicon. We start every morning with glass of this water, and try to have 3-4 more glasses during the day.

# Conclusion

During the time we devoted to write this book, many areas have forged ahead: new results have been obtained, new hypotheses developed. However, new developments are known only by a relatively small group of scientists. Despite being published in prestigious scientific journals, new research is hardly perceived by the scientific establishment. We hope that our work will make a small contribution to the dissemination of new ideas with the result of creating an active information environment.

Every year scientists working in this area participate in the international scientific Congress *"Physics, Chemistry and Biology of Water,"* held for more than 20 years under the chairmanship of Professor Gerald Pollack (www.waterconf.org). In the proceedings of the Congress, you can find articles and presentations by many scientists, whose works were reviewed in this book. Even more information can be discovered in the online journal (www.waterjournal.org).

In Russia, this subject is presented in St. Petersburg at the international scientific Congress *"Science, Information, and Spirit,"* held every year in early July (www.sis-congress.com) and the conference *"Weak and hyper-weak fields and radiations in biology and medicine"* (www.biophys.ru/arxiv/congress).

If you are interested in the development of modern science, we invite your participation within these forums.

# Index